# Drying Science and Technology

*Edited by Suvanjan Bhattacharyya*

Published in London, United Kingdom

Drying Science and Technology
http://dx.doi.org/10.5772/intechopen.105258
Edited by Suvanjan Bhattacharyya

Contributors
Laura Garcia-Llobodanin, Alejandra Billiris, Sachin Namdeo Kothawade, Vishal Pande, Vaibhav Wagh, Kalyani Autade, Sandesh Bole, Rajashri Sumbe, Shubhangi Albhar, Ildikó Nagy, Natalía Reynisdóttir, Páll Thor Ingvarsson, Arnór Freyr Ingunnarson, Ásta María Einarsdóttir, Sendhil Kumar Natarajan, Guna Muthuvairavan, Saranya Manikandan, Elavarasan Elangovan, Peter A Obasa, Bolanle Adenike Adejumo, James Agajo, Samuel Tunde Olorunsogo, Labake Fadipe, Rogelio Moreno-Muñoz, Gregorio Antolín-Giraldo, Alejandro Reyes-Salinas



First published in London, United Kingdom, 2024 by IntechOpen
IntechOpen is the global imprint of INTECHOPEN LIMITED, registered in England and Wales, registration number: 11086078, 5 Princes Gate Court, London, SW7 2QJ, United Kingdom

British Library Cataloguing-in-Publication Data
A catalogue record for this book is available from the British Library

Additional hard and PDF copies can be obtained from orders@intechopen.com

Drying Science and Technology
Edited by Suvanjan Bhattacharyya
p. cm.
Print ISBN 978-1-83769-210-1
Online ISBN 978-1-83769-211-8
eBook (PDF) ISBN 978-1-83769-212-5

For EU product safety concerns:
IN TECH d.o.o., Prolaz Marije Krucifikse Kozulić 3, 51000 Rijeka, Croatia,
info@intechopen.com or visit our website at intechopen.com.

# Meet the editor

Dr. Suvanjan Bhattacharyya is an assistant professor in the Department of Mechanical Engineering, Birla Institute of Technology and Science Pilani (BITS Pilani), India. He completed his post-doctoral research at the Department of Mechanical and Aeronautical Engineering, University of Pretoria, South Africa, under the supervision of Prof. Josua P. Meyer. Dr. Bhattacharyya obtained his Ph.D. in Mechanical Engineering from Jadavpur University, Kolkata, India, with the collaboration of Dusseldorf University of Applied Sciences, Germany. He received his master's degree from the Indian Institute of Engineering, Science and Technology, India (formerly known as Bengal Engineering and Science University) in heat-power engineering. His research interest lies in computational fluid dynamics in fluid flow and heat transfer, specializing in laminar, turbulent, steady, and unsteady separated flows and convective heat transfer, experimental heat transfer enhancement, solar energy, and renewable energy. He is the author and co-author of more than 100 papers in high-ranking journals and prestigious conference proceedings. Dr. Suvanjan was listed among the top 2% of scientists in the world by Stanford University, USA, in 2021, 2022, and 2023. He received a young scientist research travel fellowship to France from the French Institute in India (IFI), Embassy of France. He has received the best paper award at several international conferences as well. He is also on the editorial boards of fifteen journals and serves as a reviewer for more than seventy others. He is presently supervising twelve Ph.D. students and has several research grants to his credit from DST, DRDO, ISHRAE, and others. He has also implemented several international collaborative projects. His strong collaboration with international universities has led to research grants from different funding agencies.

# Contents

# Preface

It is with great pleasure and enthusiasm that I present this comprehensive volume, *Drying Science and Technology*. As the field of drying continues to evolve with advancements in research and technology, the need for a comprehensive resource that encapsulates the latest knowledge becomes increasingly essential. This book aims to fill that void by providing a thorough exploration of the principles, methodologies, and applications within the realm of drying.

Drying, a fundamental unit operation in various industries plays a pivotal role in the preservation, processing, and manufacturing of a wide array of materials. From food and pharmaceuticals to ceramics and textiles, understanding the intricacies of drying processes is indispensable for optimizing efficiency, quality, and sustainability.

This book is designed to cater to the needs of students, researchers, and professionals seeking a deeper understanding of drying phenomena. It encompasses a diverse range of topics, including the fundamentals of heat and mass transfer during drying, various drying methods and techniques, mathematical modeling, and the latest innovations in drying technology.

Each chapter is crafted to provide a blend of theoretical foundations and practical applications, offering readers a holistic perspective on the subject. The contributions from esteemed experts in the field aim to bridge the gap between theory and practice, making this book a valuable reference for both academia and industry.

As the editor, my goal is to foster a deeper appreciation for the science and technology of drying, stimulate further research endeavors, and contribute to the continuous improvement of drying processes across diverse sectors. I extend my heartfelt gratitude to all the contributors whose expertise has enriched this compilation.

I trust that this book will serve as a valuable resource for anyone interested in delving into the intricacies of drying science and technology. We hope that this book not only adds to the existing body of knowledge but also motivates additional research and innovation in the dynamic field of drying science.

**Dr. Suvanjan Bhattacharyya**
Department of Mechanical Engineering,
Birla Institute of Technology and Science Pilani,
Rajasthan, India

Section 1

# Renewable Energy Drying

Chapter 1

# Assessment of Solar Dryer Performance for Drying Different Food Materials: A Comprehensive Review

*Guna Muthuvairavan, Saranya Manikandan, Elavarasan Elangovan and Sendhil Kumar Natarajan*

## Abstract

Studying crucial drying parameters, such as activation energy and moisture diffusivity, offers valuable insights for optimizing food safety. Accurate predictions and simulations through mathematical thin-layer models aid in designing, controlling, and optimizing drying operations for various food items. Solar drying presents a viable and eco-friendly solution for food preservation. This chapter critically evaluates solar drying performance for various vegetables, fruits, marine products, and other commodities, providing comprehensive insights into its efficiency. According to the literature, the moisture diffusivity ($m^2/s$) for vegetables has been reported to be within the range of $2.01 \times 10^{-10}$–$1.935 \times 10^{-8}$. For fruits, the moisture diffusivity varies between $1.33 \times 10^{-10}$ and $6.98 \times 10^{-9}$. In the case of marine food products, the range is found to be $2.8 \times 10^{-8}$–$3.408 \times 10^{-7}$, while for other commodities, it falls between $1.79 \times 10^{-9}$ and $1.061 \times 10^{-7}$. The activation energy (kJ/mol) for vegetables has been observed to fall within the range of 24.81–47.19. Similarly, for fruits, the activation energy varies between 2.56 and 45.20. Notably, Ginger demonstrates an activation energy of 35.675 kJ/mol. Experimental results showed that lower activation energy and higher moisture diffusivity accelerate dehydration.

**Keywords:** solar drying, natural convection, forced convection, moisture diffusivity, activation energy, mathematical modeling

## 1. Introduction

Meeting the food demands of a rapidly rising global population is a significant concern for civilization. By 2050, the world's population is expected to reach 9.1 billion, demanding an additional 70% of the food supply. Most of this population expansion is likely attributed to emerging countries, with many currently suffering

| Food substance | Post-harvest loss (%) |
|---|---|
| Cereals | 20–40 |
| Pulse and oil crops | 10–20 |
| Roots and tubers | 10–20 |
| Vegetables and fruit | 40–50 |
| Meat | 20–30 |
| Fish | 20–30 |
| Diary | 10–20 |

**Table 1.**
*Post-harvest loss of various food substances [1, 3, 4].*

from hunger and food insecurity. Concerns about rising food consumption are exacerbated by increasing urbanization, climate change, and land use for non-food crop production. Most governments have focused their policies over the last few decades on improving agricultural production, land management, and population control to address rising food demand. However, despite being a critical issue, post-harvest loss does not receive the attention it needs, with fewer than 5% of research funds committed to it in previous years [1–3]. One-third of the world's food supply, or around 1.3 billion tons per year, is lost or wasted, as reported by the Food and Agriculture Organization of the United Nations (FAO). The post-harvest loss of various food substances are presented in the **Table 1**. These losses happen at every stage of the food system, from cultivation to processing to distribution to final consumption.

Principal causes of post-harvest losses include Insects, rodents, fungi, and bacteria that can damage or ruin food due to pests and diseases. Physical damage can occur when food is handled roughly, improperly stored, or transported. During storage and processing, nutrients can be lost from food. Food can be squandered due to spoilage, overproduction, or insufficient demand. Post-harvest losses have a substantial influence on food security. They decrease the quantity of available food, which can result in hunger and malnutrition. They also increase food costs, making it unaffordable for some people [1, 3, 4].

## 2. Drying-solution to post-harvest loss

Drying is a widely used solution to mitigate post-harvest losses in various agricultural products. The drying process involves reducing the moisture content of harvested crops, which helps prevent spoilage, inhibit microbial growth, and maintain the quality of the product for extended periods [5, 6]. However, it is essential to note that the effectiveness of drying as a solution for post-harvest losses can vary depending on various factors, such as the type of crop, climatic conditions, available drying methods, and storage facilities. Appropriate drying techniques must be selected and implemented to ensure that the quality and nutritional value of the crops are maintained during the drying process. The best drying method for a particular application will depend on several factors, including the type of food being dried, the desired quality of the dried product, and the available resources [7, 8].

## 3. Importance of solar drying

Solar energy is a renewable and clean source of power that harnesses sunlight to generate electricity or heat. Using solar panels made of semiconductors, photons from sunlight are absorbed, releasing electrons and creating an electric current. This energy is environmentally friendly, as it does not produce harmful emissions or deplete finite resources. Solar energy can be deployed at various scales, providing energy independence and resilience, particularly in remote areas. Although initial costs can be high, advancements in storage technologies are improving efficiency and overcoming limitations associated with weather conditions. Overall, solar energy offers a sustainable solution for our energy needs, contributing to a cleaner and more sustainable future [9, 10]. Solar drying methods offer several advantages compared to conventional drying techniques. Energy efficiency [8], cost-effectiveness [11], environmental sustainability [12], preservation of nutritional quality [13], and enhanced product quality [14] are the advantages of using solar dryers.

In this regard, Ekechukwu et al. conducted a comprehensive review of various solar energy drying system designs, construction details, and operational principles. Their findings indicated that properly designed forced convection (active) solar dryers are generally more effective and controllable than natural-circulation (passive) types. However, due to the need for electricity or fossil-fuel-driven fans and auxiliary heating sources, active solar dryers are unsuitable for remote rural village farm use in most developing countries, given their high capital, maintenance, and operational costs. On the other hand, for large-scale applications in rural areas, the "ventilated greenhouse dryer" offers the advantage of being cost-effective and simple to construct and operate on-site [15].

Fudholi et al. gave the technical directions for developing solar-assisted drying systems for agricultural produce [16]. Jairaj et al. reviewed solar dryers exclusively for grape drying on a normal scale. They included various pre-treatment and drying methods for good-quality grape drying [17]. For the Malaysia location, the air-based solar collectors integrated solar drying system was reviewed by Fudholi et al. They have included the energy, exergy, economic and environmental aspect of the various solar dryers [18]. Mustayenp et al. presented a study on various solar dryers' design, performance, and application. This review focused on solar dryer models suitable for producing high-quality dried products [19]. Hicham El Hage et al. extensively reviewed the economic and environmental aspects of the solar drying system. The critical parameters, such as payback period and $CO_2$ mitigation, were compared [20].

Om Prakash et al. reviewed the various modeling technics, including computational fluid dynamics (CFD), adaptive-network-based fuzzy inference system (ANFIS), artificial neural networking (ANN), FUZZY, thermal, mathematical, drying kinetic, and energy modeling [21]. Azwin Kamarulzaman et al. reviewed the global advancement of solar drying technologies and their prospects. They discussed various performance parameters, including energy assessment, payback period, and $CO_2$ mitigation [22]. Aprajeeta Jha et al. reviewed the recent advancements in design, application, and simulation Studies of hybrid solar drying technology. The review discussed the various software used for simulating the solar drying system, including PHOENICS, FLUENT (general purpose software with Multiphysics capabilities), FIDAP (modeling complex physics), ANSYS CFX, COMSOL Multiphysics, TRNSYS [23]. Nukulwar et al. focused on various materials used to construct solar dryers and their performance evaluation for agricultural products [24].

The literature shows that various authors reviewed the performance evaluation of some specific food commodities, the use of simulation software, various modeling, and economic and environmental aspects of solar drying systems. However, the performance evaluation of various ranges of fruits, vegetables, marine food products, and other commodities was not reported exclusively. This chapter mainly focused on the review of the performance of solar drying techniques for a range of food substances which is widely used in India, including vegetables (bottle gourd, carrot, potato, ivy gourd, and onion), fruits (banana, cucumber, Tomato, and grapes), marine food products (Fish, shrimp, and prawn), and other commodities (Ginger, Chili, and Jaggery). The drying parameters such as moisture diffusivity, activation energy, drying rate, operating temperature, size, and shape of the drying product were compared for mentioned food substances. A suitable mathematical thin-layer model for food substances was also presented.

## 4. Types of solar drying

Solar drying is an age-old technique that utilizes the sun's power to remove moisture from various substances, including food. This preservation method has been practiced for centuries and continues to be widely used in many regions worldwide. By harnessing solar energy, solar drying offers a natural and cost-effective way to extend the shelf life of perishable items while preserving their nutritional value. Different solar drying methods, each with its unique approach and design, are presented in **Figure 1**. The description, suitability, advantage, and disadvantages of each significant solar drying technique are presented in **Table 2**. These methods are often tailored to suit specific needs, environmental conditions, and the type of dried material.

**Figure 1.**
*Types of solar energy-based drying methods [7].*

| Type | Description | Suitability | Advantages | Disadvantages | Reference |
|---|---|---|---|---|---|
| Open sun drying | The product to be dried is simply placed in the sun, as shown in **Figure 2**. | Various drying needs | Simple and affordable. | Highly dependent on weather conditions. | [7, 11] |
| Direct solar dryers | The product to be dried is exposed to the sun directly, as illustrated in **Figure 3**. | Low-scale drying needs | Low cost, simple to build and operate. | It can only be used during the day when the sun is shining and exposure to environmental elements. | [5, 25] |
| Indirect solar dryers | The product to be dried is not exposed to the sun directly. The air that is used to dry the product is heated by the sun and then circulated around the product as represented in **Figure 4**. | Delicate/ valuable items | Controlled drying environment and Can be used at night or on cloudy days. | Slightly higher cost than direct solar dryers and is more complex to build and operate. | [11, 14, 26] |
| Mixed-mode solar dryers | A combination of direct and indirect solar drying **Figure 5**. | Improved efficiency | Combines the advantages of both direct and indirect solar dryers. | More complex to build and operate. | [14, 27] |
| Hybrid solar dryers | A solar dryer that uses an auxiliary energy source, such as a fan or a heater, to supplement the solar energy **Figure 6**. | Variable drying needs | It can be used in areas with low solar irradiance or on cloudy days. | More complex to build and operate. | [12, 23, 28, 29] |

**Table 2.**
*Comparison of different solar dryers.*

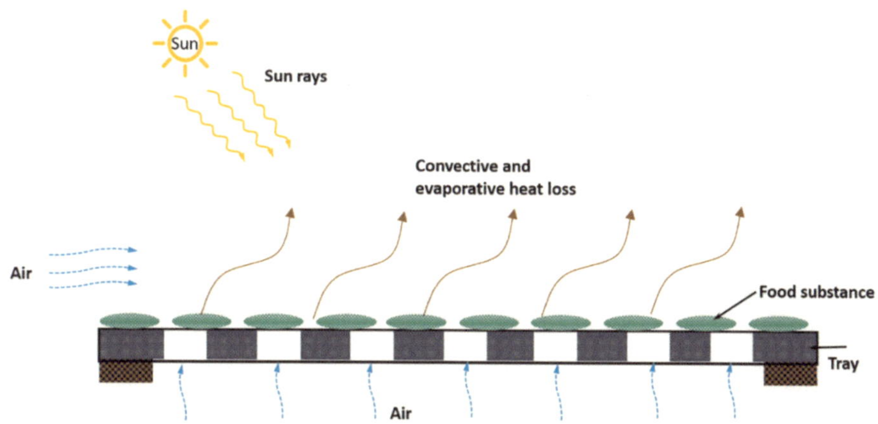


**Figure 2.**
*Schematic of open sun drying [8].*

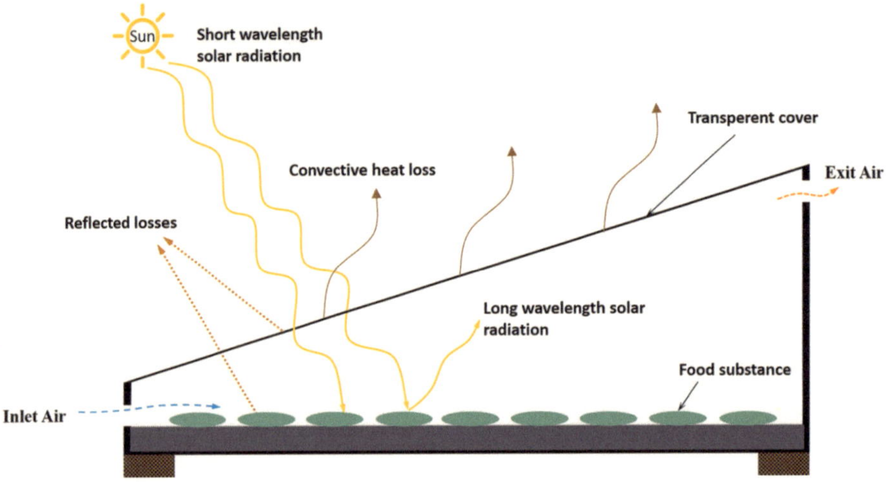


**Figure 3.**
*Schematic of the direct solar dryer [7].*

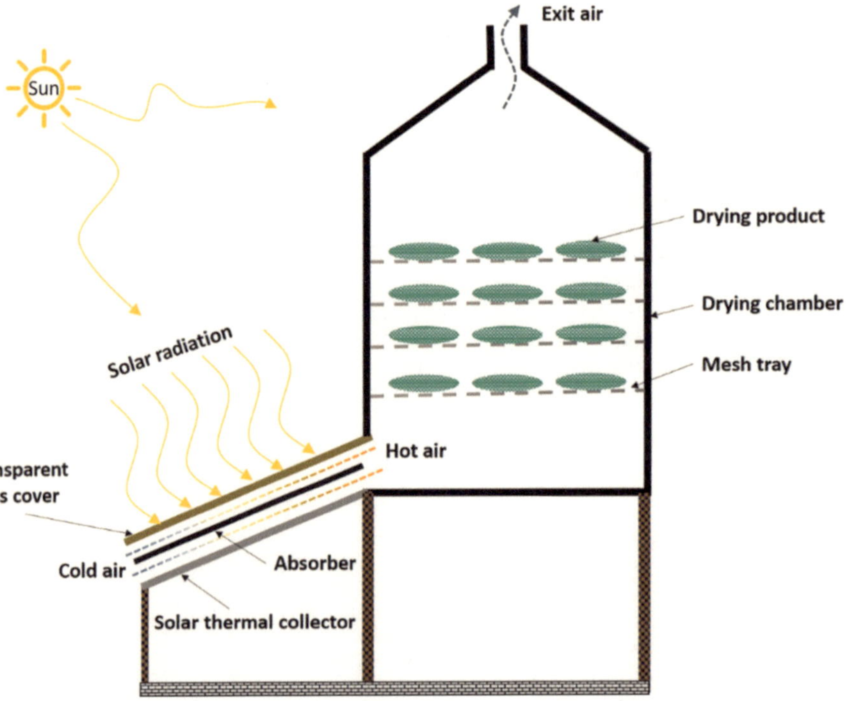


**Figure 4.**
*Schematic of the indirect solar dryer [14].*

## 5. Drying kinetics

This section presents the mathematical formulation of important drying parameters including moisture content, moisture ratio, moisture diffusivity, activation energy, and various thin-layer mathematical model.

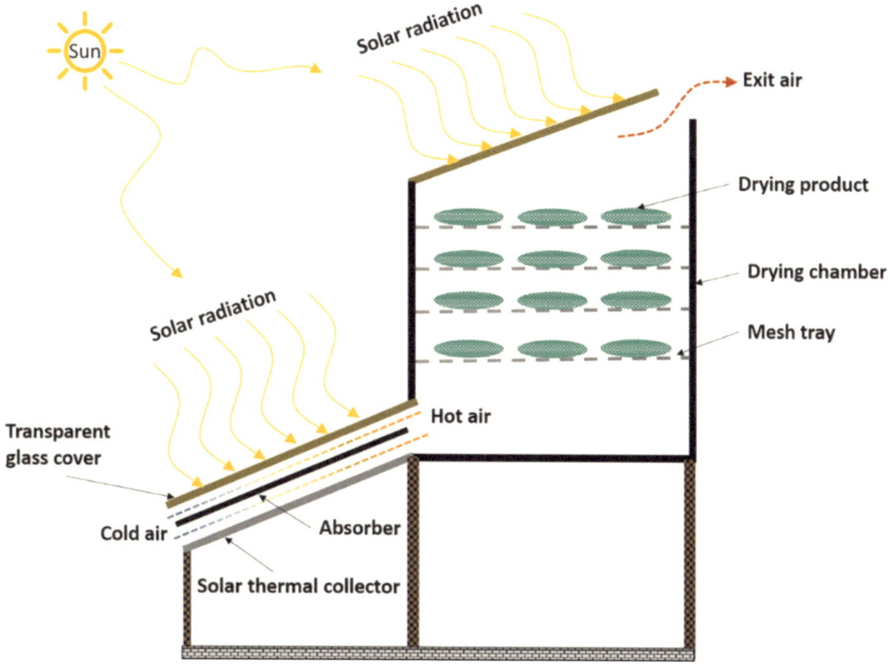


**Figure 5.**
*Schematic of the mixed-mode solar dryer [14].*

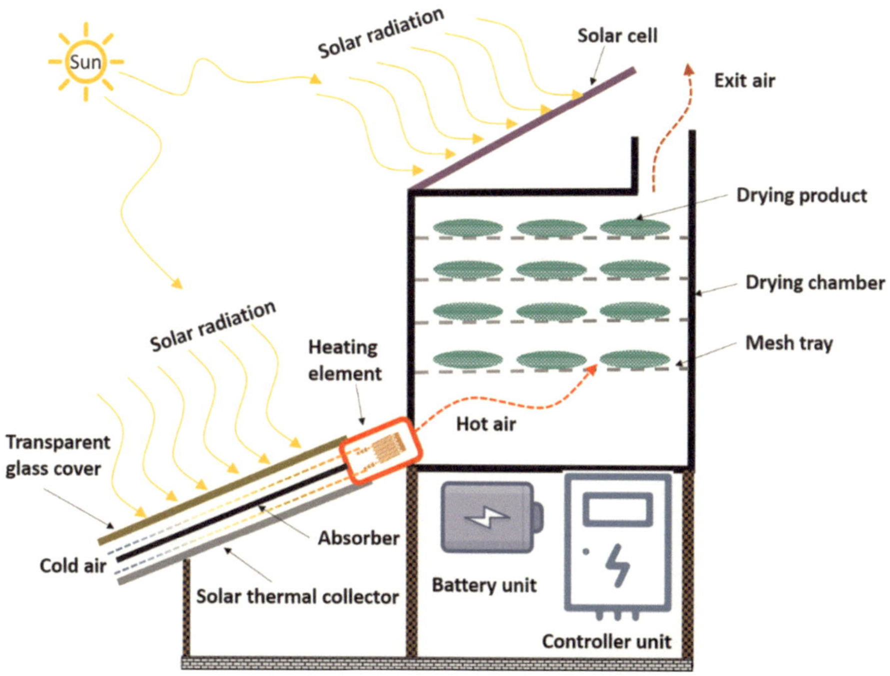


**Figure 6.**
*Schematic of the hybrid-mode solar dryer [12].*

The moisture content (MC) of a drying product refers to the amount of moisture or water present in the product, expressed as a percentage of its total weight, as presented in Eq. (1), [30].

$$MC = \frac{m_w - m_d}{m_w} \tag{1}$$

Where, $m_w$ is the mass of the drying product before drying and $m_d$ is the mass of the dried product.

Activation energy is a crucial concept in the realm of drying products, denoting the energy required to initiate and propel the drying process [31]. This context refers explicitly to the energy necessary to surmount the molecular forces that bind water molecules to the product's surface, thus enabling their evaporation. This activation energy measures the minimum energy essential for the drying process to occur at a noteworthy rate. It is important to note that the activation energy is unique to the material undergoing drying and the specific drying conditions applied. Different materials and various drying methods may exhibit distinct activation energies. Empirical investigations are typically carried out to ascertain this value, involving the study of drying kinetics for the material at different temperatures.

Another fundamental property relevant to drying processes is moisture diffusivity. This parameter describes the rate at which moisture traverses a drying product during the drying procedure [31]. Understanding moisture diffusivity is paramount in modeling and comprehending the drying kinetics of diverse materials. The moisture diffusivity factor represents the capability of water molecules to move through the product's microstructure, and its value is influenced by several factors, including temperature, humidity, and the inherent nature of the material being dried. To determine the moisture diffusivity experimentally, the moisture content is measured at various locations within the drying product over time. The acquired data is then fitted into appropriate mathematical models, such as Fick's second law of diffusion, enabling the calculation of the diffusivity coefficient [13].

The diffusion mechanism governs the drying process of food substances at the rate of falling period. Fick's second law of diffusion governs effective moisture diffusion [30].

$$\ln MR = \ln\left(\frac{8}{\pi^2}\right) - \frac{\pi^2 D_{eff} t}{L^2} \tag{2}$$

MR is a moisture ratio, and it represents the ratio of the current moisture content of the product to its initial moisture content. $D_{eff}$ is an effective moisture diffusivity ($m^2/s$), t is the corresponding drying time (hrs), and L is the thickness of the drying sample (m). The slope of the $\ln MR$ with respect to time can be written as Eq. (3),

$$slope = \frac{\pi^2 D_{eff}}{L^2} \tag{3}$$

$$MR = \frac{M_t - M_e}{M_0 - M_e} \tag{4}$$

Where $M_t, M_e$, and $M_0$ are the instantaneous, equilibrium, and initial moisture content, respectively.

The diffusion of moisture during drying can be described as Eq. (5) by Fick's second law of diffusion as equation [13],

$$D_{eff} = D_0 e^{\left(-E_a/RT\right)} \tag{5}$$

Where $D_0$ is the diffusion factor (m$^2$/s), $E_a$ is the activation energy (kJ/mol), R is the universal gas constant (8.314 kJ/mol.K), and T is the temperature (K). The plot of $\ln D_{eff}$ against 1/T gives a straight line of slope k where the relation between $E_a$ and diffusivity coefficients can be defined through linear regression analysis and activation energy $E_a$ is evaluated by Eq. (6),

$$k = \frac{E_a}{R} \tag{6}$$

The thin-layer drying model is a mathematical representation that describes the drying kinetics of a material during the drying process [30]. It assumes that the drying occurs within a thin-layer on the material's surface and considers the moisture transfer from this layer to the surrounding drying medium (usually air) [32]. Several thin-layer models have been proposed over the years. The important thin-layer model is presented in **Table 3**.

## 6. Solar drying of various food substances

In this section, solar drying of various food substances such as vegetables, fruits, marine products, and others, as shown in **Figure 7**, was elaborately discussed.

### 6.1 Vegetables

This section discusses the performance of different solar dryers for drying various vegetables, including Bottle gourd, Carrot, Potato, Ivy gourd, and Onion. The summary of the same is presented in **Table 4**.

| Name | Model | Reference |
|---|---|---|
| Newton | $MR = e^{-kt}$ | [30] |
| Page | $MR = e^{-kt^n}$ | [32] |
| Modified page | $MR = e^{\left[-(kt)^n\right]}$ | [30] |
| Henderson and Pabis | $MR = ae^{(-kt)}$ | [32] |
| Logarithmic | $MR = ae^{(-kt)} + c$ | [30] |
| Midilli-Kucuk | $MR = ae^{(-kt^n)} + bt$ | [32] |
| Wang and Singh | $MR = 1 + at + bt^2$ | [32] |

**Table 3.**
*Different thin-layer models for drying.*

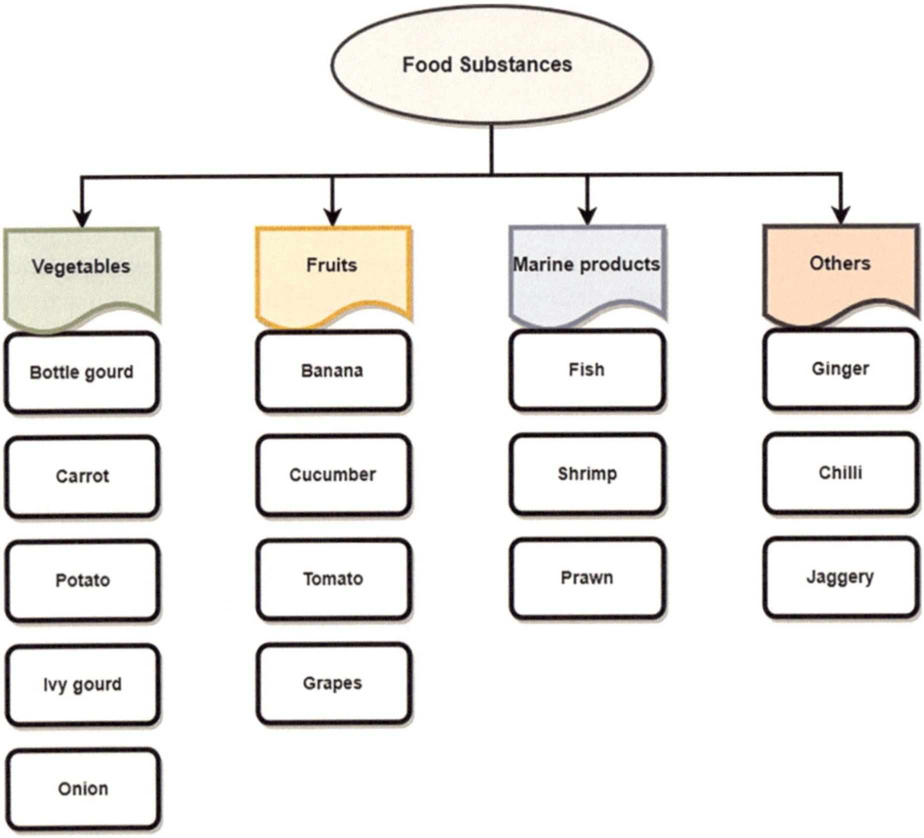


**Figure 7.**
*List of various food substances.*

### *6.1.1 Bottle gourd*

K.A. Shinde et al. evaluated the drying characteristics of steam-blanched and non-blanched bottle gourd samples. A tray dryer with forced convection and hot air was used to carry out the drying process. The samples were pulverized for analysis, and their initial moisture content was 0.04 kg water per kg dry matter. The study aimed to determine the effect of blanching on the drying efficiency and qualitative parameters of bottle gourd. The samples are pulverized for examination, and the research intends to determine the effect of blanching on the drying efficiency and qualitative features of bottle gourd [33].

Kukadiya Vishal et al. investigated using a fluidized bed dryer with forced convection for drying samples of various forms; specifically slab shapes 10 × 10 × 3 mm, 10 × 10 × 5 mm, and 10 × 10 × 7 mm. The study compared the effects of various temperatures on quality measures, including water activity and color. According to the data, the sample with dimensions of 10 × 10 × 3 mm and a temperature of 70°C has the optimum quality in terms of water activity. Furthermore, the sample with dimensions of 10 × 10 × 5 mm at 60°C exhibits outstanding color quality. The importance of shape and temperature selection in optimizing the quality features of dried items using a fluidized bed dryer was underlined in this study [34].

| Sample | Type of dryer | Mode of heat transfer | Air velocity (m/s) | Kinetic model | Pre-treatment | Sample shape / size | Initial M.C | Final M.C | Drying time | Temp. (°C) | Moisture diffusivity ($m^2/s$) | Activation energy (kJ/mol) | Ref |
|---|---|---|---|---|---|---|---|---|---|---|---|---|---|
| Bottle gourd | Tray Dryer | Forced convection | 1.5 | – | Steam blanch | Powdered | 0.04 kg water/kg d.m. | – | 345–660 mins | 50–80 | – | – | [33] |
| | Fluidized bed | Forced convection | 10 | Page model | – | Slab | – | – | 80–140 mins | 50–70 | $1.03 \times 10^{-9}$–$6.18 \times 10^{-9}$ | – | [34] |
| Carrot | Hybrid infrared power and hot air technique | Forced convection | 1 | Midilli-Kucuk model | – | Slices | 13.38 ± 0.67 (d.b.) | 0.12–0.14 (d.b.) | – | 45–75 | $2.01 \times 10^{-10}$–$12.10 \times 10^{-10}$ | | [35] |
| | Indirect solar dryer without a heat storage unit | Natural convection | – | – | – | Slices | 9.13 (d.b.) | 0.478 (d.b.) | 16 h | – | $6.7 \times 10^{-9}$ | – | [36] |
| | Indirect solar dryer with heat storage unit | Natural convection | – | – | – | Slices | 9.13 (d.b.) | 0.478 (d.b.) | 15 h | – | $7.24 \times 10^{-9}$ | – | [36] |
| Potato | Waste heat-based convective dryer | Forced convection | 1.4 ± 0.5 | Midilli-Kucuk model | – | Slices | – | – | – | 50–70 | $4.22 \times 10^{-10}$–$11.67 \times 10^{-10}$ | 47.19 | [37] |
| Ivy gourd | Indirect-type solar dryers | Natural convection | – | – | – | Slices | – | – | 16 h | 31–66 | $2.286 \times 10^{-9}$–$1.271 \times 10^{-9}$ | 39.85 | [38] |
| | | Forced convection | NA | – | – | Slices | – | – | 13 h | 32–62 | $2.286 \times 10^{-9}$–$1.935 \times 10^{-8}$ | 35.54 | [38] |
| | Single slope single basin solar dryer | Forced convection | 1 | – | Lemon juice | Slices | – | – | 7 h | 56.2 | $5.27 \times 10^{-9}$–$1.32 \times 10^{-10}$ | 26.06 | [13] |

| Sample | Type of dryer | Mode of heat transfer | Air velocity (m/s) | Kinetic model | Pre-treatment | Sample shape / size | Initial M.C | Final M.C | Drying time | Temp. (°C) | Moisture diffusivity ($m^2/s$) | Activation energy (kJ/mol) | Ref |
|---|---|---|---|---|---|---|---|---|---|---|---|---|---|
| | | | | | Honey | Slices | – | – | 8 h | 56.2 | $4.01 \times 10^{-10}$0–$2.13 \times 10^{-10}$ | 27.79 | [13] |
| | | | | | Ascorbic acid | Slices | – | – | 7 h | 56.2 | $5.85 \times 10^{-9}$–$4.47 \times 10^{-9}$ | 24.81 | [13] |
| | | | | | Sugar solution | Slices | – | – | 7 h | 56.2 | $5.26 \times 10^{-9}$–$2.71 \times 10^{-10}$ | 27.05 | [13] |
| Green onion | Solar dryer assisted by a photovoltaic module | Natural convection | – | Page and Overhults model | – | Small circles | 91.01 ± 0.15% (w.b.) | 1.64 ± 0.15 g (w.b.) | – | – | $5.15 \times 10^{-9}$ | – | [39] |
| | | Forced convection | NA | Page and Overhults model | – | Small circles | 91.46 ± 0.16% (w.b.) | 1.94 ± 0.23 (w.b.) | – | – | $1.15 \times 10^{-8}$ | – | [39] |
| Onion | Thin-layer infrared radiation drying | Forced convection | 1–1.5 | Third order polynomial | – | Slices | – | – | 7–9 h | 35–45 | $0.21 \times 10^{-1}$–$1.57 \times 10^{-10}$ | – | [40] |

**Table 4.**
*Summary of performance of different dryer for drying various vegetables.*

### *6.1.2 Carrot*

The researchers investigated drying sliced carrots utilizing hybrid infrared power, hot air drying techniques, and thermal energy storage. The carrots' diffusivity ranged from $2.01 \times 10^{-10}$ to $12.10 \times 10^{-10}$ $m^2/s$, while the specific energy used during drying ranged from 30.20 to 87.51 MJ/kg. Carrot shrinkage was measured to be between 23.49 and 51.25%. The Midilli-Kucuk model was discovered to be the best fit for the drying kinetics. The study indicated that the thermal energy storage arrangement produced promising results for future large-scale applications. These findings emphasized the possibility of adding thermal energy storage into the drying process of carrots, which would increase efficiency and quality in industrial-scale drying processes [35].

Whereas sliced carrots were dried using a passive indirect solar dryer, compared setups without (configuration 1) and with thermal energy storage (configuration 2). Configuration 1 had a drying rate of 0.5, while Configuration 2 had a greater drying rate of 0.59. In configuration 2, the sample dried faster, with the moisture content dropping from 9.13 to 0.478 on a dry basis. The average effective diffusivity for Configuration 1 was determined to be $6.7 \times 10^{-9}$ $m^2/s$ and $7.24 \times 10^{-9}$ $m^2/s$ for configuration 2. The two settings' specific energy consumption and moisture extraction rates were 3.5 and 0.28 kg/kWh for configuration 1 and 0.29 and 3.62 kg/kWh for configuration 2. Configuration 2 had a higher drying efficiency with an average of 10.25% compared to configuration 1. Based on the study's findings, it is considered that configuration 2, with the passive indirect sun dryer and thermal energy storage, is an acceptable recommendation for future large-scale applications. These findings showed the benefits of drying sliced carrots with a passive, indirect sun drier with thermal energy storage. Configuration 2 indicated higher drying efficiency, energy utilization, and a faster drying rate. This study gave valuable insights for optimizing the drying process of carrots using renewable energy sources, and it can guide future large-scale industrial applications [36].

### *6.1.3 Potato*

A waste heat-based convection dryer was used to dry potato samples. The study evaluated the drying kinetics of potato samples and found the best model to describe the process. The drying process's activation energy was 47.19 kJ/mol. This parameter gives information about the drying process's temperature sensitivity by indicating the energy required to remove moisture from potato samples. According to the study results, the Midilli et al. model offered the best fit to characterize the drying kinetics of the potato samples. This model is often used to depict removing moisture during drying. Furthermore, the effective moisture diffusivity of the potato samples was determined to be $4.22 \times 10^{-10}$–$11.67 \times 10^{-10}$ $m^2/s$ at temperatures ranging from 50 to 70°C. The effective moisture diffusivity measures moisture's ability to travel within potato samples during drying [37].

Siyabonga Gasa et al. used a solar-venturi dryer to model the drying process of sweet potato slices in naturally ventilated warm air. The drying characteristics were studied using a non-linear regression approach. The naturally ventilated solar-venturi dryer and a lemon juice pre-drying treatment were suitable for small to medium-scale drying of sweet potato slices in the study. The dryer arrangement allows for efficient drying in warm air while utilizing solar energy. The naturally-ventilated solar-venturi dryer's effective diffusivity ($D_{eff}$) values ranged from $3.32 \times 10^{-9}$ to $6.31 \times 10^{-9}$ $m^2/s$.

On the other hand, the $D_{eff}$ values for the hot air oven dryer ranged from $1.02 \times 10^{-8}$ to $2.19 \times 10^{-8}$ $m^2/s$. According to these findings, the naturally ventilated solar-venturi dryer had lower effective diffusivity values than the hot air oven drier. This implied that the solar-venturi dryer setup provides a more regulated drying environment, enhancing drying efficiency and preserving sweet potato slices [41].

### *6.1.4 Ivy gourd*

While drying ivy gourd, the researchers compared natural and forced convection indirect-type solar dryers. Various metrics were used to compare the performance of the two types of dryers. The natural convection solar dryer had an average collector efficiency of 62.56%, whereas the forced convection solar dryer had a higher average collector efficiency of 77.2%. The researchers also calculated the average values of activation energy, mass transfer coefficient, heat transfer coefficient, and diffusion coefficient for ivy gourd drying. The activation energy values of 39.85 and 35.54 kJ/mol were obtained, demonstrating the energy required for moisture elimination during drying. According to the evaluation results, the forced convection configuration produced the most significant results for drying ivy gourd. The increased drying performance was aided by better collector efficiency and favorable activation energy, mass transfer coefficient, heat transfer coefficient, and diffusion coefficient values [38].

Another study examined how various pre-treatments affected the qualitative characteristics, moisture diffusivity, and activation energy of solar-dried ivy gourd. Ascorbic acid, lemon juice, sugar solution, honey dip, and a control group were used as pre-treatments. The effective moisture diffusivity of dried ivy gourd samples varied depending on the pre-treatment. According to the findings, pre-treatments substantially impact solar-dried ivy gourd's moisture content and quality features. The findings show that pre-treatments can efficiently decrease moisture content and contribute to extended preservation periods. The lemon juice samples were found to be the best regarding moisture diffusivity and activation energy among the pre-treatments tested, demonstrating their effectiveness in the drying process [13].

Elavarasan Elangovan et al. investigated the drying kinetics of ivy gourd using a solar dryer, explicitly contrasting passive and active mode solar dryers with traditional sun drying. The results revealed that for the passive mode solar drier, the safe moisture content, indicating the necessary moisture level for storage, was obtained in 9 hours, for the active mode solar dryer in 7 hours, and for sun drying in 11 hours. This suggests that, compared to typical sun drying, both solar dryers were more efficient in drying time. The rate of moisture evaporation from the ivy gourd is affected by air temperature, whereas relative humidity influences the moisture content of the surrounding air. Airspeed, or air movement, aids in the removal of moisture-laden air from the drying environment, allowing for faster drying. The study suggested optimizing and managing certain drying parameters might improve the drying process, resulting in improved drying kinetics. Altering and regulating the solar drier's air temperature, relative humidity, global radiation, and airspeed might produce more efficient and faster ivy gourd drying [42].

Elavarasan Elangovan et al. conducted an experimental investigation to determine the investigated convective and evaporative heat transfer coefficients during the drying process of ivy gourd using natural and forced convection solar dryers and open sun drying. For ivy gourd, the average evaporative heat transfer coefficient, representing the efficiency of moisture evaporation, ranged from 181.89 to 421.84 $W/m^2$ °C.

The rate at which heat is delivered to the product and utilized for evaporation is indicated by this coefficient. Higher air velocity, as achieved by forced convection, resulted in a faster drying rate for the ivy gourd samples. Increased air velocity accelerates heat and mass transmission, resulting in faster drying. Furthermore, as drying air velocity rose, the mass transfer coefficient, representing the moisture transfer rate from the ivy gourd to the drying air, increased. This suggests that higher air velocity improves more efficient moisture removal from ivy gourd samples [25].

#### *6.1.5 Onion*

G.P. Sharma et al. dried onion slices using a thin-layer infrared radiation drying technique. The effective moisture diffusivity, which represents the rate of moisture movement within onion slices, ranged from $0.21 \times 10^{-10}$ to $1.57 \times 10^{-10}$ $m^2/s$. The drying process used forced convection, which involves the utilization of air circulation to improve heat and moisture transfer. This method allows for faster drying and increases the drying system's efficiency. Furthermore, the drying time was calculated concerning the amount of infrared power used during drying. The drying time was shortened by nearly 2.25 times when the infrared power was increased from 300 to 500 W. This suggests that the drying of the onion slices was expedited by increasing infrared power. A third-order polynomial relationship was discovered to correlate several elements influencing the specified drying process. This relationship aided in predicting drying behavior based on variables, including infrared power, drying duration, and moisture content [40].

Hidalgo et al. dried green onions using a direct sun drier aided by a photovoltaic module, focusing on natural and forced air convection operation. The effective diffusivity values for natural convection were $5.15 \times 10^{-9}$ $m^2/s$ and $1.15 \times 10^{-8}$ $m^2/s$ for forced convection. These values indicate the rate of moisture diffusion within the green onions during drying. The Page and Overhults models were selected as the best-fit models during the slower drying periods [39].

### 6.2 Fruits

This section discusses the performance of different solar dryers for drying various fruits, including Banana, Cucumber, Tomato, and Grapes. The summary of the same is presented in **Table 5**.

#### *6.2.1 Grapes*

Traditional grape drying methods have several disadvantages, such as mass losses and low quality. A joint German-Greek research program developed low-cost solar grape dryers to address these challenges. Solar dryers use the sun's energy to heat air, which is then circulated through the dryer to dry the grapes. This drying method has several advantages over traditional methods, including reduced drying time, improved quality, and prevention of mass losses [47].

The drying kinetics of two varieties of grapes grown on both shores of the Mediterranean Sea was the subject of another study. The drying kinetics were evaluated as a function of drying conditions, and the diffusion coefficient was determined. Two diffusion models were employed to determine the effective diffusivity: a simplified model based on Fick's law and a more complex model that accounted for the grapes' shrinkage. The study revealed that the drying kinetics of the two grape varieties were

| Sample | Type of dryer | Mode of heat transfer | Air velocity (m/s) | Kinetic model | Sample shape/size | Initial M.C | Final M.C | Drying time | Temp. (°C) | Moisture diffusivity (m2/s) | Activation energy (kJ/mol) | Ref |
|---|---|---|---|---|---|---|---|---|---|---|---|---|
| Red Banana | Direct solar dryer | Natural convection | – | Newly developed model | Slices | 76% (w.b.) | 18% (w.b.) | 10 h | 71.3 | $0.87–1.56 \times 10^{-9}$ | 22.56–35.49 | [42] |
| | Direct solar dryer | Forced convection | NA | Newly developed model | Slices | 76% (w.b.) | 18% (w.b.) | 8 h | 83 | $0.87–1.56 \times 10^{-9}$ | 24.58–45.20 | |
| | Open sun drying | Natural convection | – | Newly developed model | Slices | 76% (w.b.) | 18% (w.b.) | 13 h | – | – | – | |
| Poovan Banana | Direct-type solar dryer | Natural convection | – | Newly developed model | Slices | 74% (w.b.) | 20% (w.b.) | 7 h | 51 | $1.6–2.5 \times 10^{-10}$ | 23–30 | [43] |
| | Open sun drying | Natural convection | – | Newly developed model | Slices | 74% (w.b.) | 20% (w.b.) | 12 h | 66 | $1.6–2.5 \times 10^{-10}$ | 24–30 | |
| Cucumber | Double slope solar dryer | Natural convection | – | – | Slices | 93.93% (w.b.) | 11.96% (w.b.) | 3.5 days | 35–55 | – | – | [44] |
| Tomato | Hot air solar dryer | Forced convection | 0.5–2 | Page model | Slices | 93.6% (w.b.) | – | – | – | $1.58 \times 10^{-9}$– $6.98 \times 10^{-9}$ | – | [45] |
| | Hybrid dryer | Natural convection | – | Page model | Slices | 94.22% (w.b.) | 10% (w.b.) | 300 min | – | $2.00 \times 10^{-1}$– $5.84 \times 10^{-10}$ | – | [46] |
| | Solar dryer | Natural convection | – | Page model | Slices | 94.22% (w.b.) | 10% (w.b.) | 360 min | – | $1.37 \times 10^{-10}$– $4.40 \times 10^{-10}$ | – | |
| | Open sun dryer | Natural convection | – | Page model | Slices | 94.22% (w.b.) | 10% (w. b.) | 420 min | – | $1.33 \times 10^{-10}$– $4.01 \times 10^{-10}$ | – | |

**Table 5.**
*Summary of performance of different dryers for drying various fruits.*

comparable, but the grapes cultivated on the southern side of the Mediterranean Sea had a higher effective diffusivity. The study also demonstrated that the drying conditions influenced the drying kinetics, with a faster drying rate at higher temperatures and reduced relative humidities [48].

Fadhel et al. compared the three solar processes to dry Sultanine grapes: natural convection solar drier, tunnel greenhouse, and open sun. The results showed that the solar tunnel greenhouse drying was the most efficient, followed by the natural convection solar drier and open sun. The solar tunnel greenhouse drying was also the most consistent, with the drying rate being relatively unaffected by changes in weather conditions [49].

#### *6.2.2 Tomato*

J. B. Hussein et al. dried tomato slices in thin-layers using hybrid, solar, and open sun drying methods. The effective moisture diffusivity values, which represent the moisture transfer rate within the tomato slices, were determined for each drying process. The effective moisture diffusivity values in the hybrid drying method ranged from $2.00 \times 10^{-10}$ to $5.84 \times 10^{-10}$ $m^2/s$, according to the results. The values for solar drying ranged from $1.37 \times 10^{-10}$ to $4.40 \times 10^{-10}$ $m^2/s$, whereas open sun drying ranged from $1.33 \times 10^{-10}$ to $4.01 \times 10^{-10}$ $m^2/s$. On a wet basis, the moisture content of the tomato slices was reduced by 94.22–10% following drying. The Page model was used to simulate the drying kinetics. The declining rate stage of drying is described by this model, in which the moisture removal rate lowers as the moisture content falls. The Page model was used to forecast drying behavior and estimate drying time for tomato slices [46].

H. Samimi, Akhijani et al. concentrated on hot air solar drying of tomato slices using forced convection. The moisture diffusivity, representing the moisture transfer rate within tomato slices, was examined at various air velocities and slice thicknesses. At an air velocity of 2 m/s and a slice thickness of 7 mm, the most significant moisture diffusivity value achieved was $6.98 \times 10^{-9}$ $m^2/s$. Higher air velocity and thicker slices facilitate faster moisture transfer during drying. At an air velocity of 0.5 m/s and a slice thickness of 3 mm, the minimum moisture diffusivity value achieved was $1.58 \times 10^{-9}$ $m^2/s$. This shows that slower moisture transfer is caused by decreased air velocity and thinner slices. The Page model was used to analyze the drying kinetics, and it provided the best fit for the experimental data [45]. P. Rajkumar et al. examined vacuum-assisted solar drying of tomato slices using a vacuum-assisted solar drier. Compared to open sun drying, the vacuum-assisted solar drying approach required less drying time for the slices. The Page model, which best fits the experimental data, was used to analyze the drying kinetics [50].

### 6.3 Marine products

This section discusses different solar dryers' performance for drying marine food products, including Fish, Shrimp, and Prawn. The summary of the same is presented in **Table 6**.

#### *6.3.1 Fish*

Pranav Mehta et al. dried fish using a mixed-mode tent-type solar drier. The Fish's moisture content was reduced from an initial value of 89% to a final value of 10%.

| Sample | Type of dryer | Mode of heat transfer | Air velocity (m/s) | Kinetic model | Initial M.C | Final M.C | Drying time | Average temperature (°C) | Moisture diffusivity ($m^2/s$) | Ref |
|---|---|---|---|---|---|---|---|---|---|---|
| Fish | Microwave heating | Forced convection | 1 | Midilli model | 2.76 (d.b.) | 0.01 (d.b.) | – | – | $7.158 \times 10^{-8}$– $3.408 \times 10^{-7}$ | [51] |
| | Tent-type solar dryer | Forced convection | 0.28 | Lewis model | 89% (w.b.) | 10% (w.b.) | 18 h | 60–65 | $1.53 \times 10^{-7}$ | [52] |
| Shrimp | Electric Dryer | Forced convection | 0.8 ± 0.04 | Midilli model | 73–79% (w.b.) | 8–10% (w.b.) | 8 h | 55 | – | [53] |
| | Oven | Forced convection | NA | Alibas model | 4.8824 kg of water/kg of dry matter | 0.35 ± 0.143 kg of water/kg of dry matter | 210–330 min | 60–80 | $2.8 \times 10^{-8}$ | [54] |
| | Vacuum | Forced convection | NA | Midilli & Kucuk | 4.8824 kg of water/kg of dry matter | 0.35 ± 0.143 kg of water/kg of dry matter | 110–190 min | 60–80 | $5.49 \times 10^{-8}$ | |
| Prawn | Direct-type solar dryer | Natural convection | – | Newly developed model | 3.621 (d.b.) | 0.081 (d.b.) | 14 h | 35.91 | – | [55] |
| Chelwa fish | Direct-type solar dryer | Natural convection | – | Newly developed model | 2.676 (d.b.) | 0.138 (d.b.) | 21 h | 39.65 | – | |
| Silver jewfish | Hybrid solar dryer | Forced convection | NA | – | 64% (w.b.) | 10% (w.b.) | 8 h | 50 | – | [56] |

**Table 6.**
*Summary of performance of different dryers for drying various marine food products.*

During the drying process, the effective moisture diffusivity was $1.53 \times 10^{-7}$ $m^2/s$. The drying kinetics were described using the Lewis model of drying. The Lewis model is widely used to study moisture transfer in porous materials during drying.

Furthermore, the study concluded that, under loaded conditions, recirculating the outlet's hot air after absorbing moisture is the most efficient energy utilization [52]. The researchers used the Page equation to predict the drying process of Fish. The drying rate of the fish samples was found to be fastest in the beginning and gradually decreased over time. The average effective diffusivity ranged from $7.158 \times 10^8$ to $3.408 \times 10^7$ $m^2/s$, demonstrating that moisture could diffuse throughout the Fish at different rates during the drying process. The moisture content of the fish samples was reduced significantly in the study, from 2.76 to 0.01 on a dry basis, suggesting the efficiency of microwave heating in drying the Fish. The Page equation was used to simulate the drying process using non-linear regression analysis, which offered a good fit for defining the drying characteristics of the fish samples [51].

#### *6.3.2 Shrimp and prawn*

D.S. Aniesrani Delefiya et al.'s studies provided valuable insights into the drying process of shrimp in an electric dryer. The initial moisture level of the shrimp samples ranged from 73 to 79% (wb) in the study on the drying characteristics of shrimp in an electric dryer, and the drying process aimed to reduce it to a final moisture content of 8–10% (wb). Many mathematical models were explored to understand and model the drying behavior of the shrimp. The Midilli model was chosen as the best-fit model for understanding the drying kinetics of shrimp [53]. The effect of several drying processes on the physical and qualitative parameters of dried shrimps was examined. The prawns were dried in an oven at 60, 70, and 80°C for 330–210 minutes and in a vacuum oven for 190–110 minutes. The usage of a vacuum pump decreased the drying time. The drying kinetics of prawns were investigated, and both techniques' appropriate moisture diffusion and activation energy were estimated.

The Alibas and Midilli and Kucuk models offered the best experimental data with a high coefficient of determination ($R^2$) for the oven and vacuum oven approaches. The final dried goods' color features, heavy metal levels, and protein analyzes were investigated. The rehydration ratio of dehydrated shrimps was also established. The study's findings revealed that the drying conditions influenced the color characteristics of the shrimps. Shrimp dried in ovens and Hoover ovens had higher brightness and yellowness scores but lower redness levels. The Pb, As, Cd, Hg, Cu, Zn, and Fe concentrations in dried prawns were below permissible levels [54].

### **6.4 Other food substances**

In addition to the above-mentioned vegetables, fruits, and marine food products, the other food substances are also undergoing post-harvest loss predominantly. This section discusses the performance of different solar dryers for drying various essential food products, including Chili, Ginger, and Jaggery. The summary of the same is presented in **Table 7**.

#### *6.4.1 Chili*

Zakaria Hossain et al. developed a solar dryer for drying chilies. The initial moisture level of the chilies placed in the drier was 73%, which decreased to 14% during

| Sample | Type of dryer | Mode of heat transfer | Air flow rate | Kinetics model | Initial M.C | Final M.C | Drying time | Average temperature (°C) | Moisture diffusivity ($m^2/s$) | Activation energy (kJ/mol) | Ref |
|---|---|---|---|---|---|---|---|---|---|---|---|
| Ginger | Solar drying | Natural convection | _ | Page model | 621.5 (d.b.) | 12.19 (d.b.) | 8 h | $57 \pm 8.5$ | $1.789 \times 10^{-9}$ | _ | [57] |
| Cassumunar ginger | Solar greenhouse dryer | Natural convection | _ | System of partial differential equations | 90% (w.b.) | 10% (w.b.) | 24 h | 30–55 | _ | _ | [58] |
| Ginger | Thin-Layer Vacuum Dryer | Natural convection | _ | Two-term model | _ | _ | _ | 40–65 | $1.859 \times 10^{-8}$–$4.777 \times 10^{-8}$ | 35.675 | [59] |
| Chili | Solar cabinet dryer coupled with gravel bed he | Forced convection | 58 $m^3$/h | _ | 88.5% (w.b.) | 7.3% (w.b.) | 56 h | 25–55 | _ | _ | [60] |
| | Chimney solar dryer with sea pebbles | Natural convection | _ | Simple linear regression | 73.12% | 7.15% (w.b.) | 36 h | _ | $1.061 \times 10^{-7}$ | _ | [61] |
| | Chimney solar dryer without sea pebbles | Natural convection | _ | Simple linear regression | 73.12% | 9.67% (w.b.) | 37 h | _ | $8.93 \times 10^{-8}$ | _ | |
| | Solar Dryer | Forced convection | 0.11 $m^3$/s | _ | 73% (w.b.) | 14% (w.b.) | 41–46 h | 44.28 | _ | _ | [62] |
| | Open sun drying | Natural convection | _ | _ | 73% (w.b.) | 18% (w.b.) | 91 h | 34.34 | _ | _ | |

**Table 7.**
*Summary of performance of different dryers for drying various other food products.*

drying. One interesting observation was that the drying rate of the chiles on the upper tray was faster than on the lower tray. This disparity in drying rates can be attributable to various factors, including heat distribution, air circulation, and direct solar exposure. The dryer uses forced convection technology, which uses a fan or blower to increase movement and speed up drying. Forced convection improves heat transfer and moisture elimination from the chilies, resulting in faster drying [62]. Francis Kumi and Bram Parbi used an innovative approach to improve the heating system and drying performance in their study on a solar chimney dryer for chili peppers. They increased the efficiency of the solar drier by including sea pebbles in the collection base.

Using a basic regression analysis model, the researchers tested the solar chimney drier's drying performance. The chili peppers had an initial moisture content of 73.12% (w.b.) that decreased to 7.15% (w.b.) after drying. The inclusion of sea pebbles in the collection base was critical in improving the solar dryer's heating mechanism. The sea pebbles absorbed and held the sun's heat, contributing to higher temperatures within the drying chamber. As a result, the drying performance improved, and the moisture removal from the chili peppers was hastened [61]. A.K. Kamble et al. used a solar cabinet drier with a gravel bed and forced convection for drying chiles. The solar cabinet drier used forced convection to improve heat and moisture transfer by boosting airflow within the drying chamber. This forced convection technique helped the chiles dry more quickly.

Furthermore, a heat storage system was added to the drying process. This technique made heat available even after sunset, allowing the drying process to continue for an additional 4 hours. The heat storage system assisted in maintaining the required temperature within the dryer, ensuring successful drying even when solar radiation was limited [60].

#### *6.4.2 Ginger*

The performance of the solar greenhouse drier for drying Ginger was examined by Nimnuan et al. The purpose of the drying method was to bring the moisture content of the Ginger down from its initial value of 90% (wb) to a final value of 10% (wb). A thin-layer drying model was used to characterize drying kinetics. The effectiveness of the solar greenhouse drier in facilitating the drying process and attaining the target moisture content was evaluated in this study. The results provide insight into the solar greenhouse dryer's capability and efficacy for drying ginger [58]. Another study on the solar drying of Ginger reported an effective moisture diffusivity of $1.789 \times 10^9\ m^2/s$, which quantifies the rate at which moisture flows within the Ginger during drying. Various mathematical models were tested to understand the drying kinetics and model the drying process. The Page model was determined to be the best fit for characterizing the drying kinetics of Ginger in the solar dryer using natural convection among the models tested. It estimates the material's drying characteristics by considering the moisture content, drying duration, and drying rate [57].

#### *6.4.3 Jaggery*

Om Prakash et al. investigated the use of fuzzy logic to predict the rate of Jaggery's moisture evaporation in a controlled environment. MATLAB software generated the fuzzy logic model, which was then validated using experimental data. The results demonstrated that the fuzzy logic model could predict the moisture evaporation rate with no more than 0.27% error. The model can be extended to various locations under varying weather conditions: ambient temperature, solar radiation, and relative humidity [63]. The objective of another study was to construct an adaptive-network-based fuzzy inference system (ANFIS) model to predict the jaggery temperature, greenhouse air temperature, and moisture evaporation during the natural convection drying of jaggery within a greenhouse. For complete drying, distinct experiments were conducted for 0.75 kg and 2.0 kg jaggery pieces measuring $0.03 \times 0.03 \times 0.01\ m^3$. The jaggery was desiccated in a roofed, even-span greenhouse with a $1.20 \times 0.78\ m^2$ floor area. MATLAB software was used to construct the ANFIS model for calculating jaggery temperature, greenhouse air temperature, and moisture evaporation. The model was also utilized to predict the greenhouse's thermal performance based on solar intensity and ambient temperature. Analytical and experimental results for jaggery drying were in excellent agreement following experimental validation of the model [64].

Using an artificial neural network (ANN), Om Prakash et al. predicted the hourly mass of jaggery during drying in a greenhouse dehydrator with natural convection. Jaggery was dehydrated until its mass fluctuated constantly. The input parameters for the ANN model were solar radiation, ambient temperature, and relative humidity. The outcomes of the ANN model were validated using experimental data on the dehydration of jaggery mass. The statistical parameters root mean square error (RMSE) and correlation coefficient ($R^2$) was utilized to determine the difference between the values predicted by the ANN model and those observed in the experimental investigation [65].

Kumar et al. developed a thermal model that could forecast the jaggery temperature, greenhouse air temperature, and moisture evaporated (jaggery mass during drying) during natural convection drying of jaggery. The jaggery was dried in a roof-type even-span greenhouse with a $1.20 \times 0.78\ m^2$ floor area. In MATLAB software, a computer program was developed to calculate the jaggery temperature, greenhouse air temperature, and moisture evaporated. The program was also utilized to forecast the greenhouse's thermal performance based on sun intensity and ambient temperature. The program was experimentally evaluated, and the findings revealed that the analytical and experimental results for jaggery drying agreed well [66].

## 7. Conclusions and prospects

In conclusion, drying is a valuable approach to reducing post-harvest losses in agriculture, offering numerous benefits such as lower moisture content, extended shelf life, and inhibition of microbial growth. By choosing appropriate drying techniques tailored to specific crops, quality objectives, and available resources, successful post-harvest preservation can be achieved, supporting sustainable agricultural practices. Based on the literature review, the following conclusions can be drawn:

- Among the various drying methods, solar drying is a sustainable and cost-effective solution for preserving agricultural products while maintaining their nutritional quality and enhancing market value. Solar dryers, including open sun drying, direct and indirect solar dryers, and mixed-mode and hybrid-mode solar dryers, harness the sun's energy efficiently, reducing reliance on non-renewable energy sources and contributing to environmental sustainability.

- Solar drying has been extensively studied for different food substances, including vegetables, fruits, marine products, and other essential items like chili, Ginger, and Jaggery. Researchers have evaluated various parameters, such as effective moisture diffusivity, activation energy, collector efficiency, and drying kinetics, to optimize the drying process and improve quality. Pre-treatments have been explored to enhance moisture removal and preservation periods for some food products.

- Studies on solar drying of vegetables revealed the benefits of forced convection, fluidized bed, and passive indirect sun dryers with thermal energy storage, highlighting improved drying efficiency and reduced drying time. For fruits, solar tunnel greenhouse, hybrid, and vacuum-assisted solar drying demonstrated advantages such as reduced drying time and improved quality. Solar drying also proved effective for marine products, achieving improved drying performance using solar chimney dryers and electric dryers.

- The moisture diffusivity values for the studied vegetables range from $0.21 \times 10^{-10}$ to $1.15 \times 10^{-8}$ ($m^2/s$), and the activation energy varies from 24.81 to 47.19 (kJ/mol).

- The moisture diffusivity values for the studied fruits range from $1.6 \times 10^{-10}$ to $6.98 \times 10^{-9}$ ($m^2/s$), and the activation energy varies from 22 to 58 (kJ/mol).

- The moisture diffusivity values for the studied marine food products range from $2.8 \times 10^{-8}$ to $1.53 \times 10^{-7}$ ($m^2/s$).

- The moisture diffusivity values for the studied other food products range from $1.789 \times 10^{-9}$ to $1.061 \times 10^{-7}$ ($m^2/s$).

- Higher moisture diffusivity and lower activation energy accelerates the dehydration process.

Solar drying presents a diverse range of methods to efficiently remove moisture from various substances while harnessing the sun's energy. It offers a sustainable and effective solution to preserve agricultural products, minimize post-harvest losses, and reduce food waste. By incorporating solar drying into agricultural practices, we can promote greener and more resilient food processing, supporting a cleaner and sustainable future for food preservation worldwide.

Future research directions should focus on further optimizing solar drying techniques, exploring innovative designs and materials, and integrating advanced control systems to enhance the performance and versatility of solar dryers. Additionally, investigations into solar drying technologies' economic feasibility and scalability are warranted to facilitate widespread adoption in both small-scale and large-scale food processing operations. Future research on modeling drying kinetics of food substances could focus on developing novel mathematical models, incorporating multi-scale modeling approaches, applying artificial intelligence and machine learning techniques, studying non-conventional drying methods, investigating coupled phenomena, analyzing quality attributes, and integrating sustainability considerations. These efforts aim to improve the accuracy and efficiency of drying models, account for complex interactions during drying, predict drying behavior under different conditions, optimize energy consumption, and minimize environmental impact. Such research will contribute to advancements in drying technologies and enhance the understanding of drying processes in the food industry.

## Acknowledgments

The authors would like to thank the Department of Mechanical Engineering at the National Institute of Technology Puducherry, Karaikal, for their support and research facilities.

## Conflict of interest

The authors declare no conflict of interest.

## Nomenclature

| Symbol | |
|---|---|
| $MR$ | moisture ratio |
| $MC$ | moisture content |
| $m_w$ | mass of the drying product before drying (kg) |
| $m_d$ | mass of the dried product (kg) |
| $M_t$ | moisture content at any time 't' (kg) |
| $M_e$ | equilibrium moisture content (kg) |
| $M_0$ | initial moisture content (kg) |
| $k$ | drying constant ($min^{-1}$) |
| $t$ | drying time (min) |
| $n$ | empirical exponent |
| $a$&$b$ | empirical parameter |
| $D_{eff}$ | effective moisture diffusivity ($m^2/s$) |
| $L$ | thickness of the drying sample (m) |
| $D_0$ | diffusion factor ($m^2/s$) |
| $E_a$ | activation energy (kJ/mol) |
| $R$ | Universal gas constant (8.314 kJ/mol.K) |
| $T$ | temperature (K) |
| $k$ | slope of $\ln D_{eff}$ against 1/T |

## Author details

Guna Muthuvairavan[1], Saranya Manikandan[1], Elavarasan Elangovan[2] and Sendhil Kumar Natarajan[1]*

1 Department of Mechanical Engineering, National Institute of Technology Puducherry, Karaikal, Union Territory of Puducherry, India

2 Department of Aerospace Engineering, SJC Institute of Technology, Chikkaballapur, Karnataka, India

*Address all correspondence to: sendhil80@nitpy.ac.in

## References

[1] Kitinoja L, Saran S, Roy SK, Kader AA. Post-harvest technology for developing countries: Challenges and opportunities in research, outreach and advocacy. Journal of the Science of Food and Agriculture. 2011;**91**:597-603. DOI: 10.1002/jsfa.4295

[2] Godfray HCJ, Beddington JR, Crute IR, Haddad L, Lawrence D, Muir JF, et al. Food security: The challenge of feeding 9 billion people. Science (80-). 2010;**327**:812-818. DOI: 10.1126/science.1185383

[3] Kumar D, Kalita D. Reducing post-harvest losses during storage of grain crops to strengthen food security in developing countries. In: Christopher JS, editor. Foods. Vol. 6. 2017. pp. 1-22. DOI: 10.3390/foods6010008

[4] Hanneman LJ. Dried fruit. In: Patisserie. London, UK: Taylor & Francis; 2013. p. 6. DOI: 10.4324/9780080940908-5

[5] Natarajan SK, Elangovan E, Elavarasan RM, Balaraman A, Sundaram S. Review on solar dryers for drying fish, fruits, and vegetables. Environmental Science and Pollution Research. 2022;**29**: 40478-40506. DOI: 10.1007/s11356-022-19714-w

[6] Natarajan SK, Elavarasan E. Experimental investigation of drying potato for Karaikal climatic condition. IOP Conference Series: Earth and Environmental Science. 2019;**312**: 1-7. DOI: 10.1088/1755-1315/312/1/012021

[7] Sharma A, Chen CR, Vu Lan N. Solar-energy drying systems: A review. Renewable and Sustainable Energy Reviews. 2009;**13**:1185-1210. DOI: 10.1016/j.rser.2008.08.015

[8] Tiwari S, Tiwari GN, Al-Helal IM. Development and recent trends in greenhouse dryer: A review. Renewable and Sustainable Energy Reviews. 2016; **65**:1048-1064. DOI: 10.1016/j.rser.2016.07.070

[9] Arjun Singh K, Muthuvairavan G, Natarajan SK. Numerical investigation of modified conical cavity receiver with different heat transfer fluids. Energy Sources, Part A: Recovery, Utilization and Environmental Effects. 2023;**45**: 6964-6980. DOI: 10.1080/15567036.2023.2218833

[10] Muthuvairavan G, Natarajan SK. Large-scale solar desalination system. In: A. T.V, V. Selvaraj, M. M. M, editors. In: Solar Thermal Conversion Technologies for Industrial Process Heating. 1st ed. Boca Raton: Taylor & Francis; 2023. pp. 169-199. DOI: 10.1201/9781003263326-9

[11] Kumar M, Sansaniwal SK, Khatak P. Progress in solar dryers for drying various commodities. Renewable and Sustainable Energy Reviews. 2016;**55**: 346-360. DOI: 10.1016/j.rser.2015.10.158

[12] Parhizi Z, Karami H, Golpour I, Kaveh M, Szymanek M, Blanco-Marigorta AM, et al. Modeling and optimization of energy and exergy parameters of a hybrid-solar dryer for basil leaf drying using RSM. Sustainability. 2022;**14**:1-27. DOI: 10.3390/su14148839

[13] Elangovan E, Natarajan SK. Effects of pre-treatments on quality attributes, moisture diffusivity, and activation energy of solar dried ivy gourd. Journal of Food Process Engineering. 2021;**44**: 1-14. DOI: 10.1111/jfpe.13653

[14] EL-Mesery HS, EL-Seesy AI, Hu Z, Li Y. Recent developments in solar drying technology of food and

agricultural products: A review. Renewable and Sustainable Energy Reviews. 2022;**157**:112070. DOI: 10.1016/j.rser.2021.112070

[15] Ekechukwu OV, Norton B. Review of solar-energy drying systems II: An overview of solar drying technology. Energy Conversion and Management. 1999;**40**:615-655. DOI: 10.1016/S0196-8904(98)00093-4

[16] Fudholi A, Sopian K, Ruslan MH, Alghoul MA, Sulaiman MY. Review of solar dryers for agricultural and marine products. Renewable and Sustainable Energy Reviews. 2010;**14**:1-30. DOI: 10.1016/j.rser.2009.07.032

[17] Jairaj KS, Singh SP, Srikant K. A review of solar dryers developed for grape drying. Solar Energy. 2009;**83**:1698-1712. DOI: 10.1016/j.solener.2009.06.008

[18] Fudholi A, Sopian K, Bakhtyar B, Gabbasa M, Othman MY, Ruslan MH. Review of solar drying systems with air based solar collectors in Malaysia. Renewable and Sustainable Energy Reviews. 2015;**51**:1191-1204. DOI: 10.1016/j.rser.2015.07.026

[19] Mustayen AGMB, Mekhilef S, Saidur R. Performance study of different solar dryers: A review. Renewable and Sustainable Energy Reviews. 2014;**34**:463-470. DOI: 10.1016/j.rser.2014.03.020

[20] El Hage H, Herez A, Ramadan M, Bazzi H, Khaled M. An investigation on solar drying: A review with economic and environmental assessment. Energy. 2018;**157**:815-829. DOI: 10.1016/j.energy.2018.05.197

[21] Prakash O, Laguri V, Pandey A, Kumar A, Kumar A. Review on various modelling techniques for the solar dryers. Renewable and Sustainable Energy Reviews. 2016;**62**:396-417. DOI: 10.1016/j.rser.2016.04.028

[22] Kamarulzaman A, Hasanuzzaman M, Rahim NA. Global advancement of solar drying technologies and its future prospects: A review. Solar Energy. 2021;**221**:559-582. DOI: 10.1016/j.solener.2021.04.056

[23] Jha A, Tripathy PP. Recent advancements in design, application, and simulation studies of hybrid solar drying technology. Food Engineering Reviews. 2021;**13**:375-410. DOI: 10.1007/s12393-020-09223-2

[24] Nukulwar MR, Tungikar VB. A review on performance evaluation of solar dryer and its material for drying agricultural products. Materials Today Proceedings. 2021;**46**:345-349. DOI: 10.1016/j.matpr.2020.08.354

[25] Elangovan E, Natarajan SK. Convective and evaporative heat transfer coefficients during drying of ivy gourd under natural and forced convection solar dryer. Environmental Science and Pollution Research. 2023;**30**:10469-10483. DOI: 10.1007/s11356-022-22865-5

[26] Lingayat AB, Chandramohan VP, Raju VRK, Meda V. A review on indirect type solar dryers for agricultural crops–Dryer setup, its performance, energy storage and important highlights. Applied Energy. 2020;**258**:114005. DOI: 10.1016/j.apenergy.2019.114005

[27] Elkhadraoui A, Kooli S, Hamdi I, Farhat A. Experimental investigation and economic evaluation of a new mixed-mode solar greenhouse dryer for drying of red pepper and grape. Renewable Energy. 2015;**77**:1-8. DOI: 10.1016/j.renene.2014.11.090

[28] Karami H, Kaveh M, Golpour I, Khalife E, Rusinek R, Dobrzański B, et al. Thermodynamic evaluation of the forced convective hybrid-solar dryer during drying process of rosemary (Rosmarinus officinalis l.) leaves. Energies. 2021;**14**: 5835. DOI: 10.3390/en14185835

[29] Ndukwu MC, Simo-Tagne M, Abam FI, Onwuka OS, Prince S, Bennamoun L. Exergetic sustainability and economic analysis of hybrid solar-biomass dryer integrated with copper tubing as heat exchanger. Heliyon. 2020; **6**:e03401. DOI: 10.1016/j.heliyon.2020.e03401

[30] Elangovan E, Natarajan SK. Experimental research of drying characteristic of red banana in a single slope direct solar dryer based on natural and forced convection. Food Technology and Biotechnology. 2021;**59**:137-146. DOI: 10.17113/ftb.59.02.21.6876

[31] Elangovan E, Natarajan SK. Effects of pre-treatments on quality attributes, moisture diffusivity, and activation energy of solar dried ivy gourd. Journal of Food Process Engineering. 2021;**44**: e13653. DOI: 10.1111/jfpe.13653

[32] Elangovan E, Natarajan SK. Experimental study on drying kinetics of ivy gourd using solar dryer. Journal of Food Process Engineering. 2021;**44**: e13714. DOI: 10.1111/jfpe.13714

[33] Shinde KA, Scholar MT, Process A, Naik V. Study of drying characteristics of bottle gourd in tray dryer. International Journal of Research Studies in Agricultural Sciences. 2016;**2**:1-7. DOI: 10.20431/2455-6224.0202001

[34] Vishal K. Drying behavior and mathematical modelling of bottle gourd. International Journal of Agriculture Environment and Biotechnology. 2019; **12**:299-306. DOI: 10.30954/0974-1712.12.2019.1

[35] Geng Z, Torki M, Kaveh M, Beigi M, Yang X. Characteristics and multi-objective optimization of carrot dehydration in a hybrid infrared/hot air dryer. Lebensmittel-Wissenschaft & Technologie. 2022;**172**:114229. DOI: 10.1016/j.lwt.2022.114229

[36] Gilago MC, Mugi VR, Chandramohan VP. Performance assessment of passive indirect solar dryer comparing without and with heat storage unit by investigating the drying kinetics of carrot. Energy Nexus. 2023;**9**: 100178. DOI: 10.1016/j.nexus.2023.100178

[37] Masud MH, Himel HH, Arefin AME, Ananno AA, Rashid M, Dabnichki P. Mathematical modelling and exergo-environmental analysis of drying potato samples in a waste heat-based convective dryer. Environmental Challenges. 2021; **5**:2-11. DOI: 10.1016/j.envc.2021.100372

[38] Gilago MC, Chandramohan VP. Performance evaluation of natural and forced convection indirect type solar dryers during drying ivy gourd: An experimental study. Renewable Energy. 2022;**182**:934-945. DOI: 10.1016/j.renene.2021.11.038

[39] Hidalgo LF, Candido MN, Nishioka K, Freire JT, Vieira GNA. Natural and forced air convection operation in a direct solar dryer assisted by photovoltaic module for drying of green onion. Solar Energy. 2021;**220**: 24-34. DOI: 10.1016/j.solener.2021.02.061

[40] Sharma GP, Verma RC, Pathare PB. Thin-layer infrared radiation drying of onion slices. Journal of Food Engineering. 2005;**67**:361-366. DOI: 10.1016/j.jfoodeng.2004.05.002

[41] Gasa S, Sibanda S, Workneh TS, Laing M, Kassim A. Thin-layer modelling of sweet potato slices drying under naturally-ventilated warm air by solar-venturi dryer. Heliyon. 2022;**8**: e08949. DOI: 10.1016/j.heliyon.2022.e08949

[42] Elangovan E, Natarajan SK. Experimental study on drying kinetics of ivy gourd using solar dryer. Journal of Food Process Engineering. 2021;**44**:1-39. DOI: 10.1111/jfpe.13714

[43] Bhanu AS, Elavarasan E, Natarajan SK, Anandu A, Senin HM. Experimental investigation of drying kinetics of poovan banana under forced convection solar drying. In: Acharya SK, Mishra DP, editors. Current Advances in Mechanical Engineering. Singapore: Springer Singapore; 2021. pp. 621-631

[44] Elavarasan E, Natarajan SK, Bhanu AS, Anandu A, Senin MH. Experimental investigation of drying cucumber in a double slope solar dryer under natural convection and open sun drying. In: Palanisamy M, Natarajan SK, Jayaraj S, Sivalingam M, editors. Innovations in Energy, Power and Thermal Engineering. Singapore: Springer Singapore; 2022. pp. 41-52

[45] Akhijani HS, Arabhosseini A, Kianmehr MH. Effective moisture diffusivity during hot air solar drying of tomato slices. Research in Agricultural Engineering. 2016;**62**:15-23. DOI: 10.17221/33/2014-RAE

[46] Hussein JB, Filli KB, Oke MO. Thin layer modelling of hybrid, solar and open sun drying of tomato slices. Research Journal of Food Science and Nutrition. 2016;**1**:15-27. DOI: 10.31248/rjfsn2016.010

[47] Mujumdar AS. Instructions for preparation of ivianuscripts for direct reproduction. Drying Technology. 1991; **9**:1353-1354. DOI: 10.1080/07373939108916764

[48] Azzouz S, Guizani A, Jomaa W, Belghith A. Moisture diffusivity and drying kinetic equation of convective drying of grapes. Journal of Food Engineering. 2002;**55**:323-330. DOI: 10.1016/S0260-8774(02)00109-7

[49] Fadhel A, Kooli S, Farhat A, Bellghith A. Study of the solar drying of grapes by three different processes. Desalination. 2005;**185**:535-541. DOI: 10.1016/j.desal.2005.05.012

[50] Rajkumar P, Kulanthaisami S, Raghavan GSV, Gariépy Y, Orsat V. Drying kinetics of tomato slices in vacuum assisted solar and open sun drying methods. Drying Technology. 2007;**25**:1349-1357. DOI: 10.1080/07373930701438931

[51] Darvishi H, Azadbakht M, Rezaeiasl A, Farhang A. Drying characteristics of sardine fish dried with microwave heating. Journal of the Saudi Society of Agricultural Sciences. 2013;**12**:121-127. DOI: 10.1016/j.jssas.2012.09.002

[52] Mehta P, Samaddar S, Patel P, Markam B, Maiti S. Design and performance analysis of a mixed mode tent-type solar dryer for fish-drying in coastal areas. Solar Energy. 2018;**170**: 671-681. DOI: 10.1016/j.solener.2018.05.095

[53] Delfiya DSA, Murali S, Alfiya PV, Samuel MP. Drying characteristics of shrimp (Metapenaeus dobsoni) in electrical dryer. Pantnagar Journal of Research. 2020;**18**:280-284. Available from: https://www.researchgate.net/publication/357735675_Drying_characteristics_of_shrimp_in_electrical_dryer

[54] Ersan AC, Tugrul N. The drying kinetics and characteristics of shrimp dried by conventional methods: Scientific paper. Chemical Industry & Chemical Engineering Quarterly. 2022; **27**:319-328. DOI: 10.2298/CICEQ201114050E

[55] Jain D, Pathare PB. Study the drying kinetics of open sun drying of fish. Journal of Food Engineering. 2007;**78**: 1315-1319. DOI: 10.1016/j.jfoodeng.2005.12.044

[56] Fudholi A, Bakhtyar B, Saleh H, Ruslan MH, Othman MY, Sopian K. Drying of salted silver jewfish in a hybrid solar drying system and under open sun: Modeling and performance analyses. International Journal of Green Energy. 2016;**13**:1135-1144. DOI: 10.1080/15435075.2016.1175347

[57] Deshmukh AW, Varma MN, Yoo CK, Wasewar KL. Investigation of solar drying of ginger (Zingiber officinale): Emprical modelling, drying characteristics, and quality study. Chinese Journal of Engineering. 2014; **2014**:1-7. DOI: 10.1155/2014/305823

[58] Nimnuan P, Nabnean S. Experimental and simulated investigations of the performance of the solar greenhouse dryer for drying cassumunar ginger (Zingiber cassumunar Roxb.). Case Studies in Thermal Engineering. 2020;**22**:100745. DOI: 10.1016/j.csite.2020.100745

[59] Thorat ID, Mohapatra D, Sutar RF, Kapdi SS, Jagtap DD. Mathematical modeling and experimental study on thin-layer vacuum drying of ginger (Zingiber Officinale R.) slices. Food and Bioprocess Technology. 2012;**5**:1379-1383. DOI: 10.1007/s11947-010-0429-y

[60] Kamble AK, Pardeshi IL, Singh PL, Ade GS. Drying of chilli using solar cabinet dryer coupled with gravel bed heat storage system. Journal of Food Science and Technology. 2017;**1**:87-94

[61] Kumi F, Parbi B, Amoah RS, Ampah J. Computational fluid dynamics and performance evaluation of a chimney solar dryer for chilli pepper. Agricultural Engineering International: CIGR Journal. 2022;**24**:153-165

[62] Hossain M, Hossain MZ, Hossain MA, Awal MA, Alam MM, Rabbani AHMM. Design and development of solar dryer for chilli drying. International Journal of Research. 2015;**2**:63-78. Available from: http://edupediapublications.org/journals/index.php/ijr/article/view/1292

[63] Prakash O, Kumar A, Kaviti AK, Kumar PV. Prediction of the rate of moisture evaporation from jaggery in greenhouse drying using the fuzzy logic. Heat Transfer Research. 2015;**46**: 923-935. DOI: 10.1615/HeatTransRes.2015007463

[64] Prakash O, Kumar A. ANFIS modelling of a natural convection greenhouse drying system for jaggery: An experimental validation. International Journal of Sustainable Energy. 2014;**33**:316-335. DOI: 10.1080/14786451.2012.724070

[65] Prakash O, Kumar A. Application of artificial neural network for the prediction of jaggery mass during drying inside the natural convection greenhouse dryer. International Journal of Ambient Energy. 2014;**35**:186-192. DOI: 10.1080/01430750.2013.793455

[66] Kumar A, Tiwari GN. Thermal modeling of a natural convection greenhouse drying system for jaggery: An experimental validation. Solar Energy. 2006;**80**:1135-1144. DOI: 10.1016/j.solener.2005.09.011

Chapter 2

# Experimental Investigation on Drying of Forest Biomass Particles in a Mechanically Stirred Fluidized Bed

*Rogelio Moreno-Muñoz, Gregorio Antolín-Giraldo and Alejandro Reyes-Salinas*

## Abstract

This chapter reflects a review of the results obtained in several investigations on the fluidization process of wet biomass particles in a mechanically stirred fluidized bed equipment. For the experimental purposes, an experimental equipment with a drying column of 300 mm in diameter is used. Using the Ergun equation, the fluodynamic behavior of the bed is analyzed to obtain, from the measurements of pressure drops in the bed and imposed velocities, the specific gas-particle contact surface; such interfacial surface varies between 5758 and 7317 $m^2$ $m^{-3}$ for dry particles and between 2774 and 4444 $m^2$ $m^{-3}$ for high humidity particles. Subsequently, the phenomena of heat and mass transfer by convection between the fluidizing gas and the biomass particles during the drying process are studied. The gas-particle heat and mass transfer coefficients are determined, considering stirrer rotation speeds between 1 and 2 rev $s^{-1}$. The convective coefficients vary between 13 and 25.7 W $m^{-2}$ $K^{-1}$ for heat transfer and between 6 x $10^{-3}$ and 20 x $10^{-3}$ m $s^{-1}$ for mass transfer; thus, correlations have been obtained between the Nusselt and Reynolds numbers and between the Sherwood and Reynolds numbers, respectively, valid in the Reynolds number range between 102 and 257.

**Keywords:** fluidization, drying, heat and mass transfer, biomass, particle drying, agitated fluidized bed

## 1. Introduction

Forests and the wood industry's residues represent an extraordinary source of energy for enterprises and communities located near areas where wood is harvested, and wood byproducts are produced. Currently, there is a clear conviction about the need to make rational use of this energy resource. Therefore, interest in the development of renewable energy studies will continue in the coming years. In the case of biofuels, it is a significant field for applied research.

Among the alternatives for using these resources, there is the possibility of combustion or gasification to generate thermal and/or electrical energy, or the option of producing fertilizer products. The process of densifying wood from forest residues or the wood industry can also be an important alternative due to its low density, low calorific value, and in some cases, high moisture content, which results in high biomass consumption. In industrial processes, the reuse of waste for transformation into byproducts with higher added value, such as particleboard or wood fibers, also generates environmental benefits.

In thermochemical processes such as the combustion of forest biomass, the limitation is the low calorific value of the fuel due to its high moisture content. For this reason, burning them without prior drying requires the use of large volumes of combustion chambers to process the fuel. In this sense, a drying process is important as a preliminary stage to combustion to reduce the volume of the combustion chamber, achieve an adequate combustion temperature, and improve the thermal efficiency of the process.

On the other hand, biomass gasification and pyrolysis processes are normally carried out with low humidity. Thus, for example, according to [1, 2] sawdust gasification in a fluidized bed requires moisture contents of 8–12% and 8.5%, respectively.

In the field of particle board manufacturing, briquettes or pellets require very low levels of humidity. Indeed, the transformation of green wood into particles is done with moisture contents that can vary between 35 and 150% d.b., but the subsequent processes of gluing and pressing the particles require that the moisture be between 5 and 10% to the outer layers and from 2 to 6% for the middle layer. In this context [3], carried out a study on the drying of wood particles in rotary equipment with a humidity of 65% to leave the biomass with a final moisture content of 9–11% d.b., thus allowing hot pressing of the particles for the manufacture of MDF boards. Zabaniotou [4] also studied the drying of forest biomass in rotary equipment but with low initial moisture levels (13% d.b.).

It can then be seen that in all the industrial applications mentioned, the humidity values required for the various transformation processes of the residual biomass are well below the humidity with which it is normally available and that can be 150–170% d.b. It is not always possible to have these residues in a dry state and among the preparatory treatments that are required is the drying of the biomass, which has motivated this research.

Fluidized bed technology has become one of the most successful worldwide. It currently finds applications not only in thermal processes such as combustion, gasification and drying, but also in others on roasters, calciners, classifiers and reactors within the metallurgical, chemical, and pharmaceutical sectors.

In a previous work [5] the authors carried out a thermal study on the drying of biomass particles in a mechanically stirred fluidized bed equipment; the objective of the study was to determine the production rates of forest biomass particles, the evaporation rates, the specific energy consumption, and the thermal efficiency of the drying process. The most important results of this study reveal that the specific energy consumption is 3040 kJ for each kg of water evaporated in the bed, which corresponds to a thermal efficiency of 80%.

In this research the first objective is to experimentally analyze the aerodynamics of the fluidized bed, in order to obtain minimum fluidization velocities, pressure drops in the bed and the specific surface of the gas-particle contact. The second objective is to study the phenomena of heat and mass transfer that occur simultaneously during the drying of the biomass particles, in order to determine the convective coefficients of heat and mass transfer in the contact surface between the biomass particles and the fluidizing gas.

## 2. Theoretical background

In solid drying processes, momentum, heat, and mass transport phenomena occur simultaneously. The study of these phenomena is important to define aspects of the mechanical and thermal design of a drying equipment. Particularly, in a fluidized bed dryer, the study of the fluid dynamics of the bed and the determination of minimum fluidization velocities and pressure drop in the bed have been the objective of many studies.

In some cases, the problem has been considered as a particle isolated from the rest of the particles; in others, the interaction with other particles is considered (a situation that always occurs in a fluidized bed). It is well known that in fluidized beds with irregular particles, turbulent flow occurs, in which case energy losses due to kinetic effects are more important than losses of viscous origin.

On the other hand, in the thermal design of a fluidized bed dryer, it is important to know the heat and mass transfer coefficients that occur at the particle-gas interface because with them the boundary conditions can be formulated in modeling problems of the drying process. Particularly, during the period of constant drying rate, in which the predominant drying mechanism is convective at the gas-particle interface.

### 2.1 Transport of momentum in a fluidized bed

In fluidized systems, it is necessary to perform analysis using constitutive equations, which are added to the momentum balance equations.

In the case of flows in porous matrices with low-velocity flows and low Reynolds numbers, the Darcy equation relates the resistive force between the fluid and the porous matrix with the superficial velocity of the fluid. On the other hand, for fluidized beds of large particles, a situation that occurs in a particulate biomass dryer, the resistant force depends on the viscous losses and the kinetic energy losses.

Therefore, in a fluidized bed, the equation that determines the pressure drops, developed by Ergun [6] establishes that:

$$\frac{\Delta p}{L} = 150\frac{(1-\varepsilon)^2}{\varepsilon^3}\frac{\mu_g U}{{D_p}^2} + 1.75\frac{1-\varepsilon}{\varepsilon^3}\frac{GU}{D_p} \tag{1}$$

In the case of fluidized beds with irregular particles (non-spherical), the Ergun equation must take into account the sphericity of the particle through the relationship $6/\phi D_p$, that represents the specific surface area of the particles $S_p$ for arbitrary particles of similar size and expressed as the contact surface of particles per unit volume of solids.

Therefore, in the case of non-spherical particles, Ergun equation can be rewritten in terms of $S_p$ as follows:

$$\frac{\Delta p}{L} = 150\frac{(1-\varepsilon)^2}{\varepsilon^3}\frac{\mu_g U}{36}{S_p}^2 + 1.75\frac{1-\varepsilon}{\varepsilon^3}\frac{GU}{6}S_p \tag{2}$$

If the superficial velocity $U$ is lower than the minimum fluidization velocity $U_{mf}$, the bed remains in its rest condition and Eq. (2) allows determining the specific surface area $S_p$ of the particles contained in the bed. For this, experimental data of the pressure drop as a function of superficial velocity must be available under the condition that $U < U_{mf}$.

### 2.2 Gas-particle convective heat and mass transfer in fluidized bed

For the analysis of heat transfer between a spherical particle and a gas with relative velocity $U$ (spherical liquid drop falling in a gas), in Ranz and Marshall [7] the following dimensionless equation is proposed:

$$Nu_{gp} = \frac{h_{gp}D_p}{k_g} = 2.0 + 0.60\left(\frac{\rho_g U D_p}{\mu_g}\right)^{1/2}\left(\frac{C_g \mu_g}{k_g}\right)^{1/3} \quad (3)$$

In the analysis of heat transfer in fluidized beds, particularly at high fluidization velocities and Reynolds numbers, there are dimensionless correlations similar to Eq. (3). For example [8] proposes the following correlation for coarse particles ($Re_p > 100$) and in a fixed bed condition:

$$Nu_{gp} = 2 + 1.8 {Pr_g}^{1/3} {Re_p}^{1/2} \quad (4)$$

For low values of Reynolds number, this equation presents important deviations in relation to experimental results. However, in such cases, the experimental results correlate well with those predicted by the Kunii and Levenspiel equation, whose fundamental assumption is the existence of a plug flow regime:

$$Nu_{gp} = 0.03 {Re_p}^{1.3} \quad (5)$$

for $0.1 < Re_p < 100$, under the assumption of a plug flow.

On the other hand, and in a similar way to what happens with the phenomenon of heat transfer, the analysis of the mass transfer in a forced convection regime between an isolated spherical particle and a gas in relative motion can be carried out using the experimental correlation of [9]:

$$Sh_{gp} = \frac{k_{gp}D_p}{D_v} = 2.0 + 0.60\left(\frac{\rho_g U D_p}{\mu_g}\right)^{1/2}\left(\frac{\mu_g}{\rho_g D_v}\right)^{1/3} \quad (6)$$

for $0.6 < Sc_g < 2.7$ and $2 < Re_p < 800$.

For fixed beds [7], also reported an expression for the calculation of the Sherwood number for large particles ($Re_p > 80$) in liquid and gas systems. When applied to gaseous systems, the equation can be written as:

$$Sh_{gp} = 2.0 + 1.8 Re_p^{1/2} Sc_g^{1/3} \quad (7)$$

In analogy with the phenomenon of heat transport, for flow regimes with low Reynolds numbers, Eq. (7) does not adequately predict the value of the Sherwood number and in these cases, it is recommended to use the following equations proposed by Richardson and Szekely [10].

$$Sh_{gp} = 0.374\, Re_p^{1.18} \text{ for } 0.1 < Re_p < 15 \tag{8}$$

$$Sh_{gp} = 2.01\, Re_p^{0.5} \text{ for } 15 < Re_p < 250 \tag{9}$$

## 3. Material and methods

### 3.1 Aerodynamics in an agitated fluidized bed of biomass particles

The experiments were carried out in a laboratory equipment whose main section, the drying chamber, is shown in **Figure 1**; details of the equipment, which has also been used to carry out the experiment on heat and mass transfer, can be found in Moreno et al. [11]. **Figure 2** shows an overview of the equipment.

To develop the aerodynamic analysis of the fluidized bed, experiments were carried out to determine the pressure drop of the bed as a function of the superficial velocity of the air. The basis weight of the biomass samples was 2.0 kg d.b. The superficial velocity was gradually increased until reaching the $U_{mf}$ value in order to analyze the bed in its rest condition; the suspended bed condition was also guaranteed, for which the superficial velocities rose above the minimum fluidization velocity.

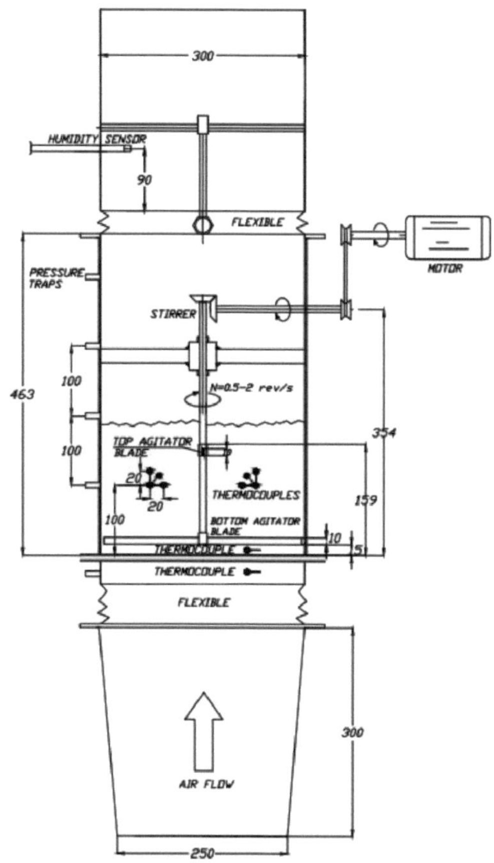


**Figure 1.**
*Schematic diagram of experimental equipment.*

**Figure 2.**
*Overview of the fluidized bed dryer of biomass particles.*

The $d_p$ particle size varied in the range of 0.51–3.56 mm. The tests were carried out first using particles with equilibrium moisture content (0.15 kg kg$^{-1}$ d.b.); subsequently, the bed was loaded with wet particles (2.0 kg kg$^{-1}$ of humidity d.b.), with which the density of the particles varied between 423 and 1044 kg m$^{-3}$.

The experimental velocity data were obtained with a Pitot tube. On the other hand, the pressure drops of the bed were measured with a differential manometer for each superficial velocity tested. In addition, the rotation speed of the mechanical agitator was varied, seeking the best stability condition for the bed of wet particles, since wet biomass particles have a great tendency toward agglomeration as a result of surface forces generated by the water contained in them.

In a fixed bed of biomass particles, the bed porosity $\varepsilon$ can be calculated with the values of the density of the particles $\rho_p$ and the density of the static bed $\rho_b$. Thus,

$$\varepsilon = 1 - \frac{\rho_b}{\rho_p} \tag{10}$$

In order to analyze the quality of the fluidization of biomass particles, prior to the drying tests, the procedure used consists of determining the fraction of particles in the bed that are suspended by the ascending air flow; the fluidization quality index *QF* is used according to the Eq. (11). Thus, if the bed is completely fluidized, the pressure drop of the bed should be equal to the weight of solids per unit of cross-sectional area of the bed and otherwise the *QF* index is less than 1.

$$QF = \frac{\Delta p}{W/A} \tag{11}$$

### 3.2 Fundamentals of gas-particle heat transfer

Regarding the determination of the convective heat transfer coefficient in fluidized beds, in most of the works reported in the literature, the interfacial heat transfer analysis is carried out in batch processes and steady state, where the hot gas enters the bed and is then cooled by making contact with the cold particles. Thus, for a bed layer of thickness d$l$, the convective coefficient can be represented as:

$$-C_g U \rho_g \frac{dT_g}{dl} = h_{gp} S\left(T_g - T_{p,s}\right) \tag{12}$$

which, integrated for a height $l$ in the bed, allows to obtain:

$$\ln \frac{T_{g,i} - T_{p,s}}{T_{g,l} - T_{p,s}} = \frac{h_{gp} S}{\rho_g U C_g} l \tag{13}$$

where $T_{g,i}$ and $T_{g,l}$ are the gas temperature at the inlet and at a height $l$ of the bed.

Representing the first member of Eq. (13) as a function of $l$, based on the experimental data, then the value of $h_{gp}$ can be obtained through the slope plotted on a semilogarithmic graph. The slope can be variable if the local coefficient presents variations depending on the position inside the bed.

Eq. (13) has been obtained on the basis of the complete mixing of gas and particles, that is, assuming that the temperature of the particles in the bed is the same at all points, except for a region of the bed close to the distributor. The above is because temperature gradients and equilibrium between gas and particles are reached in this zone, as already reported by [8, 12, 13]. This means that the gas temperature variations must be measured in the area close to the distributor and that the temperature of the particles can be considered equal to that of the gas at the outlet of the bed. Some authors consider that a thermocouple inserted in the bed provides a measurement of the temperature of the solids, which, strictly speaking, is not the case. This experimental model assumes that heat losses to the environment are negligible.

For studies of heat transfer coefficients in drying systems in the period of constant drying rate, it can be assumed that the temperature of the surface particles $T_{p,s}$ is equal to the wet bulb temperature $T_{wb}$ of the inlet air [14, 15], since in such a period the product shows a wet surface.

The analysis of the drying curves, obtained in Moreno [16] allows us to conclude that a large part of the process is carried out under a constant temperature regime and at a constant drying rate. In principle, the determination of the heat transfer coefficients should be carried out using Eq. (13). However, this procedure provides a local value of the convective heat transfer coefficient.

For design purposes, it is more appropriate to work with mean values of $h_{gp}$. Thus, in an adiabatic regime, the heat balance equation can be written as:

$$-h_{fg}\rho_{p,0}\frac{dw}{dt} = h_{gp}S_p\Delta T_{ml} \tag{14}$$

Thus, the $h_{gp}$ calculation can be carried out using the equation:

$$h_{gp} = \frac{h_{fg}\rho_{p,0}\left(-\frac{dw}{dt}\right)}{S_p\Delta T_{ml}} \tag{15}$$

In Eq. (14) the concept of logarithmic-mean temperature difference is introduced, which for its application in this drying process can be assumed that the surface temperature of the particles is equal to the temperature of the wet bulb of the air because the drying was carried out under the condition of constant rate. Thus:

$$\Delta T_{ml} = \frac{T_{g,i} - T_{g,o}}{\ln\left(\frac{T_{g,i}-T_{wb}}{T_{g,o}-T_{wb}}\right)} \tag{16}$$

### 3.3 Fundamentals of gas-particle mass transfer

In analogy with the convective phenomenon of heat transfer between a solid particle and a fluid, which is governed by Newton's cooling equation, for a drying process the mass transfer between the wet particle and the air that receives the vapor released by solids is governed by a similar equation, that is:

$$\dot{m}_v = k_{gp}A_p(c_{v,s} - c_{v,\infty}) \tag{17}$$

It can be assumed, in analogy with heat transfer, that the moisture concentration in the humid air $c_{v,\infty}$ of Eq. (17), at a given bed height, is uniform across the cross section of the drying chamber and the transfer of matter is completed in a very small distance above the distributor.

By making a mass balance during the drying process in the constant drying rate period, for a differential fluidized bed element of thickness d$l$, the mass flow of water vapor at the particle surface, based on Eq. (17), can be determined. Thus:

$$d\dot{m}_v = k_{gp}(c_{v,s} - c_v)SAdl \tag{18}$$

The water vapor flow can be related to the moist content of the moist air through:

$$d\dot{m}_v = \dot{m}_{da}dw_a \tag{19}$$

Then:

$$\dot{m}_{da}dw_a = k_{gp}(c_{v,s} - c_v)SAdl \tag{20}$$

If incompressible flow is considered, then:

$$\dot{m}_{da} v_a dc_v = k_{gp}(c_{v,s} - c_v) SAdl \tag{21}$$

When Eq. (21) is integrated from a point on the distributor, for a bed height $l$, the following is obtained:

$$\ln \frac{c_{v,s} - c_{v,i}}{c_{v,s} - c_{v,l}} = \frac{k_{gp} SA}{\dot{m}_{da} v_a} l \tag{22}$$

On the other hand, using an average coefficient of $k_{gp}$, the mass balance equation for the entire bed can be written as:

$$k_{gp} = \frac{\dot{m}_v}{A_p \Delta c_{ml}} \tag{23}$$

where the logarithmic mean difference in moisture concentration is defined as:

$$\Delta c_{ml} = \frac{c_{v,o} - c_{v,i}}{\ln\left(\frac{c_{v,s} - c_{v,i}}{c_{v,s} - c_{v,o}}\right)} \tag{24}$$

Since the equation applies to the entire bed, then $A_p = S_p V_p$ and

$$k_{gp} = \frac{\dot{m}_v}{S_p V_p \Delta c_{ml}} \tag{25}$$

In terms of the drying rate ($-\mathrm{d}w/\mathrm{d}t$) Eq. (25) can be written as:

$$k_{gp} = \frac{\rho_{p,0}\left(-\frac{dw}{dt}\right)}{S_p \Delta c_{ml}} \tag{26}$$

Similarly, to what was assumed in the heat transfer analysis, in determining the surface mass transfer coefficient it was assumed that the concentration of water vapor on the wetted surface of the particles is equal to that of saturated air and evaluated at the wet bulb temperature of humid air [15].

### 3.4 Experiments on heat and mass transfer

The experimental dryer has a drying chamber with a diameter of 0.3 m. The biomass load, as in the fluodynamic tests, was 2.0 kg d.b., occupying an approximate height of 0.17 m. The mechanical stirrer was built with a vertical shaft and four blades in total.

To analyze the homogeneity of the bed and its temperature, 8 temperature sensors were installed, as shown in **Figures 1** and **3**, and connected to a Digi-Sense Cole-Parmer Instrument Company multichannel thermometer; it has a resolution of 0.1 K and $\pm$ 0.5 K accuracy and an RS-232 output for connection to a PC. Data was collected every 4 s and analyzed using a ScanLink 2.0 software. An additional temperature sensor PT100 was placed at the entrance of the air flow to the dryer to control the operating temperature of the equipment.

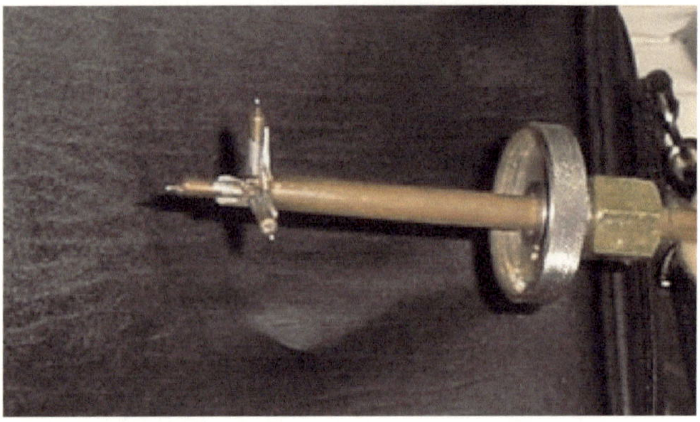

**Figure 3.**
*Arrangement of thermocouples to measure temperature inside the bed.*

At the bed outlet, a Digi-Sense digital psychrometric digital recorder was used together with ScanLink 2.0 and PCDAC (Cole-Parmer Instrument Company) programs to collect data every 10 s.

The drying experiences were carried out with Pinus radiata sawdust particles. The operating temperature measured at the inlet of the hot air flow to the drying chamber was recorded, as well as the outlet temperature of the gas above the particle bed.

In determining the mean logarithmic difference of Eq. (16) and to avoid the error caused by the heat transferred from the air to the distributor plate, the temperature of the gas at the bed inlet $T_{g,i}$ was measured at a point just above the air distributor. Thus, was verified that the air temperature, after passing through the distributor, drops sharply when it comes into contact with the particles. **Figure 4** shows a schematic diagram of the thermocouples to obtain the axial distribution of temperature in the bed. The homogenization of the temperature is reached at a height lower than 20 mm above the distributor, as shown in **Figure 5**, which agrees with the analysis of other authors [8, 12] and the experimental results of [13, 17].

The physical properties of the gas were evaluated at the mean temperature of the fluid film surrounding the particle, considering that the particle surface temperature is equal to the air wet bulb temperature.

In the heat and mass transfer experiments, three control factors ($U$, $d_p$ and $N$) were used. The experiments were carried out, considering each particle with an air velocity within the range in which its fluidization occurs according to the moisture content of the particles. In order to study possible variations of the $h_{gc}$ and $k_{gc}$ coefficients, at each level of ($d_p$-$U$) the experiments were performed with two rotation speeds of the mechanical bed stirrer. **Table 1** shows the 10 trials with their respective levels; the operating temperature was 150°C in all cases.

## 4. Results and discussion

### 4.1 Bed porosity, specific surface area, and minimum fluidization velocity

Using Eq. (10) and the procedure described in Section 3.1, the porosity of the bed has been determined in the fixed bed condition. The specific surface has been obtained based on the Ergun's modified equation. The size $d_p$ corresponds to particles

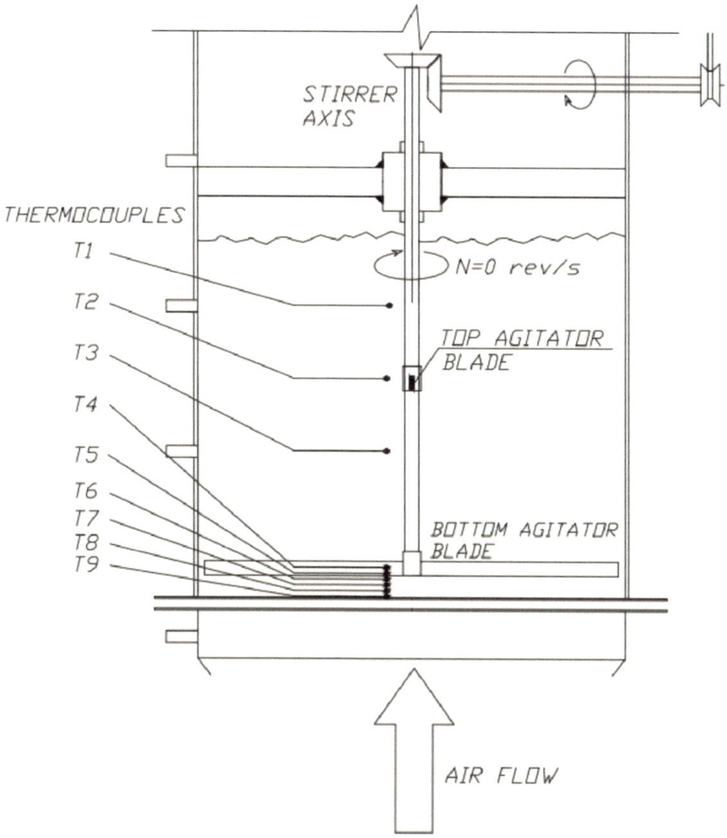


**Figure 4.**
*Schematic position of the thermocouples to determine the axial temperature distribution in the bed;* N = 0 *rev* $s^{-1}$.

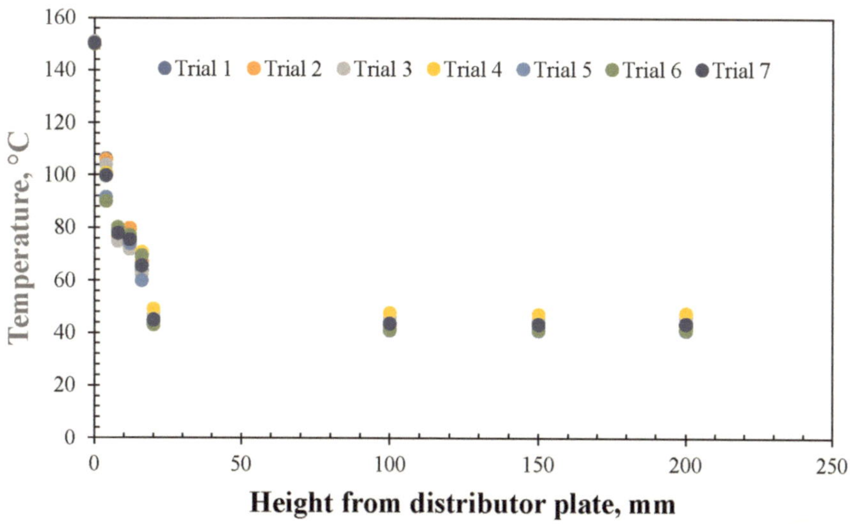


**Figure 5.**
*Axial temperature profile inside the bed, corresponding to seven profiles measured with an inlet temperature equal to 150°C.*

| $d_p$ (mm) | $U$ (m s$^{-1}$) | $N$ (rev s$^{-1}$) | $U_{mf}$ (m s$^{-1}$) |
|---|---|---|---|
| 0.89 | 0.71 | 1 | 0.46 |
| 0.89 | 0.71 | 2 | 0.46 |
| 1.44 | 0.77 | 1 | 0.56 |
| 1.44 | 0.77 | 2 | 0.56 |
| 1.85 | 0.81 | 1 | 0.66 |
| 1.85 | 0.81 | 2 | 0.66 |
| 2.18 | 0.82 | 1 | 0.80 |
| 2.18 | 0.82 | 2 | 0.80 |
| 3.56 | 1.02 | 1 | 0.95 |
| 3.56 | 1.02 | 2 | 0.95 |

**Table 1.**
*Experiments for determination of* $h_{gp}$ *and* $k_{gp}$ *coefficients.*

obtained by means of ASTM E-11 sieving. **Table 2** shows the results on bed porosity and specific surface area of particles.

In the preliminary tests in an agitated fluidized bed, the range between 0 and 2 rev s$^{-1}$ has been chosen, to analyze the behavior of the bed in terms of its mobility, depending on the combination of the superficial velocity and agitation speeds parameters. Tests were carried out with a load of 2 kg of biomass particles of 1.44 mm in size and with a moisture content of 2.0 kg kg$^{-1}$ d.b.

**Figure 6** shows the behavior of the pressure drop in the bed as a function of superficial velocity, for a stirred bed with different stirring speeds. With 0.5 rev s$^{-1}$ (**Figure 6a**) it could be seen that at a speed of 1.12 m s$^{-1}$ (point A), the bed opens abruptly giving way to a sudden increase in the air flow through the bed, reaching a velocity of 1.35 m s$^{-1}$ (point B). The fluidization stabilizes from a velocity of 1.46 m s$^{-1}$ with a suspension of the particles equivalent to a $QF$ = 0.78.

By increasing the turning speed to 1 rev s$^{-1}$ (**Figure 6b**), the transition from the fixed bed to the fluidized bed is more gradual with a transition speed of 1.18 m s$^{-1}$ (A) and a $QF$ = 0.88, but then there tends to be a decrease in $QF$ with increasing velocity. This phenomenon is significantly attenuated at higher turning speed and at 2 rev s$^{-1}$ (**Figure 6d**) at a superficial velocity of 0.91 m s$^{-1}$ (A) the transition from a fixed bed to a fluid bed occurs, reaching a $QF$ = 0.95. A notable aspect of these tests was the minimal amount of particle carryover out of the dryer.

| $d_p$ (mm) | $\varepsilon$ (m$^3$ m$^{-3}$) | $S_{p,o}$ (m$^2$ m$^{-3}$) | $S_p$ (m$^2$ m$^{-3}$) |
|---|---|---|---|
| 0.51 | 0.604 | 7317 | 4444 |
| 0.89 | 0.616 | 6835 | 3887 |
| 1.44 | 0.618 | 6440 | 3457 |
| 1.85 | 0.631 | 6243 | 3253 |
| 2.18 | 0.646 | 6118 | 3126 |
| 3.56 | 0.697 | 5758 | 2774 |

**Table 2.**
*Fixed bed porosity and specific surface area for dry and wet biomass particles.*

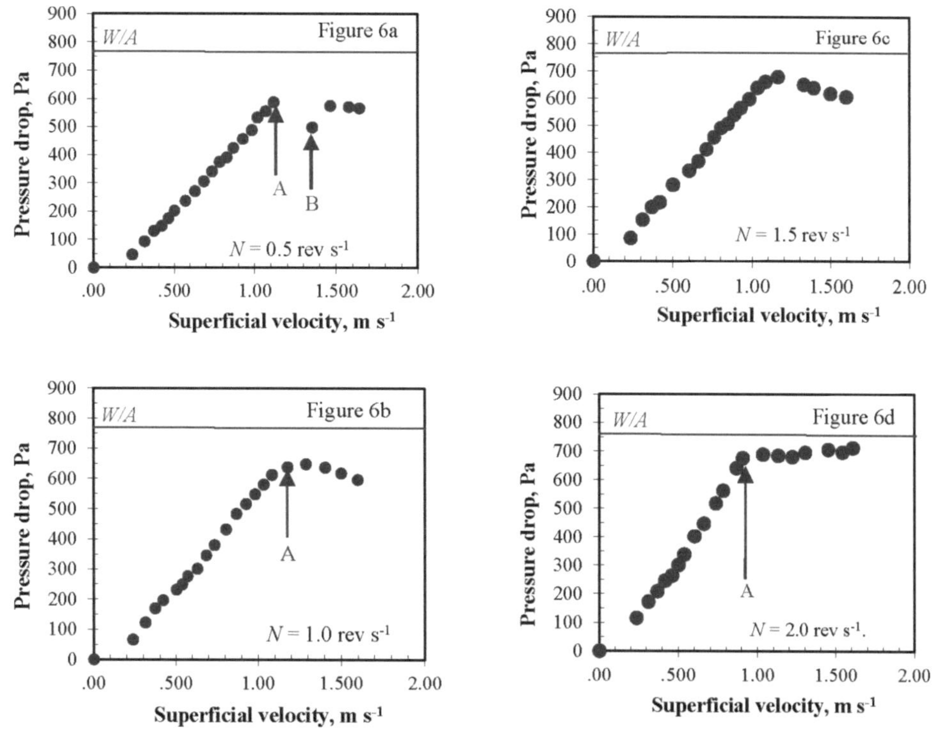


**Figure 6.**
*Pressure drop across the bed as a function of superficial velocity with different stirring speeds for* $d_p$ = 1.44 *mm and* w = 2.0 *kg kg*$^{-1}$ *d.b.*

Once the objective of having a high-quality fluidized bed was achieved, the minimum fluidization velocity was determined. **Figure 7** shows a curve of pressure drop versus superficial velocity obtained with particles with equilibrium moisture and 1.85 mm size for an agitation velocity of 2.0 rev s$^{-1}$. When carrying out tests with the other particle sizes, it was found that the qualitative behavior was similar, with obvious differences in the velocity values obtained, depending on the particle size, as shown in **Table 3**.

From the point of view of the quality of the fluidization of wet particles, it is found that mechanical agitation has an important effect when the particles have high moisture contents, taking into account that a bed of wet biomass is impossible to fluidize without mechanical agitation; in the case of low moisture particles, the effect of agitation is less.

The minimum fluidization velocities $U_{mf}$ were determined in a mechanically shaken fluidized bed with agitation velocity $N$ = 2 rev s$^{-1}$, for dry and wet biomass particles; it can be observed in **Table 3** that with dry biomass, fluidization velocities are smaller.

For the $U_{mf}$ velocity prediction, a new correlation is proposed based on a procedure reported in [18]. The initial correlation studied fits well with experimental data in the range of small particles. However, with large particle sizes, the prediction of experimental values of $U_{mf}$ was not successful and therefore the correlation had to be corrected to consider the variation of the $U_{mf}$ velocity with the particle size. To carry out this correction, a geometric factor ($d_p/D_O$) was used, where $d_p$ is the particle size obtained by sieving and $D_O$ is a reference size equal to 1 mm. Thus, the proposed correlation is:

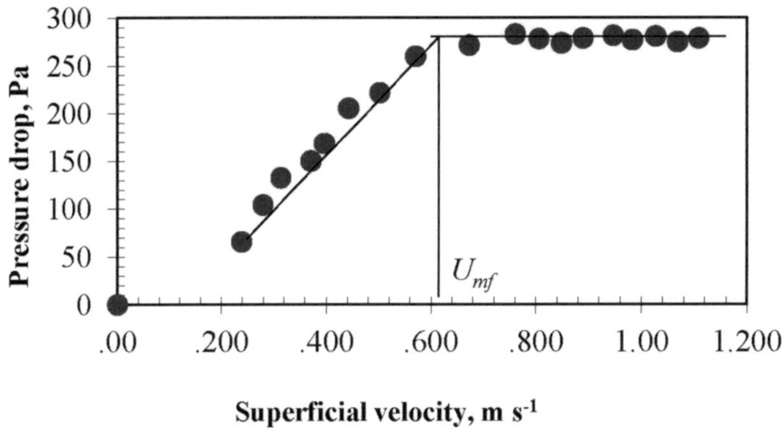


**Figure 7.**
*Pressure drop in an agitation-fluidized bed for* $d_p$ = *1.85 mm and equilibrium moisture content;* N = *2.0 rev* $s^{-1}$.

| Dry biomass | | | Wet biomass | | |
|---|---|---|---|---|---|
| $d_p$ (mm) | Experimental $U_{mf}$ | Eq. (27) | $d_p$ (mm) | Experimental $U_{mf}$ | Eq. (27) |
| 0.89 | 0.46 | 0.46 | 0.89 | — | 0.78 |
| 1.44 | 0.56 | 0.58 | 1.44 | 0.91 | 0.99 |
| 1.85 | 0.66 | 0.66 | 1.85 | 1.09 | 1.11 |
| 2.18 | 0.80 | 0.75 | 2.18 | 1.42 | 1.25 |
| 3.56 | 0.95 | 1.10 | 3.56 | 1.69 | 1.80 |

**Table 3.**
*Variation of the minimum fluidization velocity* $U_{mf}$ *in m* $s^{-1}$, *in relation to the particle size* $d_p$ *for dry and wet biomass;* N = *2.0 rev* $s^{-1}$.

$$Re_{mf} = \left[\left(24.8^2 + 0.044GaM_\rho\right)^{0.5} - 24.8\right]\left(\frac{d_p}{D_0}\right)^{1/3} \tag{27}$$

The prediction of the velocity of minimum fluidization using Eq. (27) as observed in **Table 3**, it is acceptable; only in two tests the deviations were −12% and +15%.

## 4.2 Heat and mass transfer

### *4.2.1 Thermal characterization of the bed of biomass particles*

From the point of view of the fluodynamic behavior of the particles in the fluidized bed, it is first important to guarantee the high quality of their suspension, reflected in a high value of the *QF* index of Eq. (11), as already indicated. From the thermal point of view itself, the high quality of fluidization has been evidenced in a high homogeneity of the bed temperature, according to the records of the eight thermocouples inserted inside the fluidized bed, as shown in **Figure 8**, which have been obtained during a drying experiment.

In addition, this excellent fluidization quality has been verified with the high value of the correlation coefficient $R^2$ for the drying curves, as will be seen in the drying experiments. **Figure 9** shows one of the drying curves obtained and it can be seen that the process is carried out with a constant drying rate, therefore the ideal adjustment curve in this case would be a straight line with a high value for the explanatory variance $R^2$.

On the other hand, in order to know the behavior of the biomass during the drying process, specifically in the range of temperatures normally used in dryers with atmospheric air, a thermogravimetric analysis has been carried out.

The experimental tests were carried out in a Cahn 2000 thermogravimetric equipment, 113× system, equipped with a programmable temperature control system.

**Figure 10** shows the TG diagram in an inert atmosphere, for two tests carried out with wet and dry biomass. This diagram shows the percentage of residual weight, with respect to the initial weight loaded, as a function of the sample during the test.

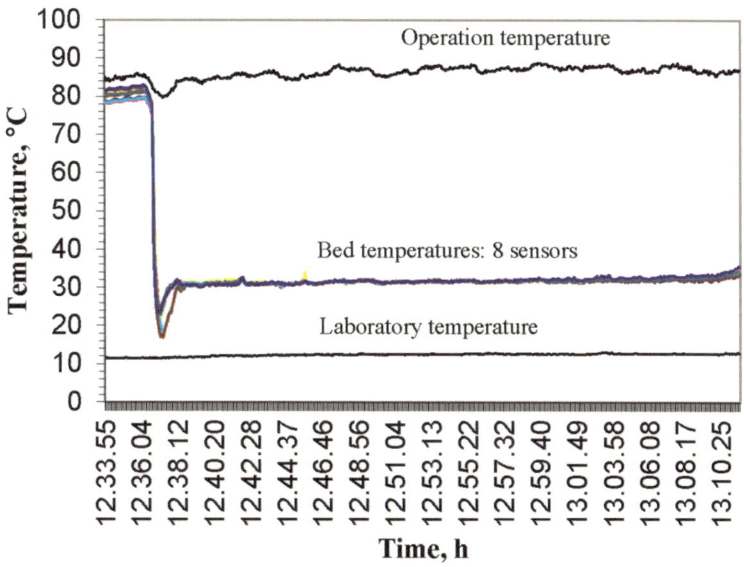


**Figure 8.**
*Temperature profile at different points in the bed during biomass drying experiment.*

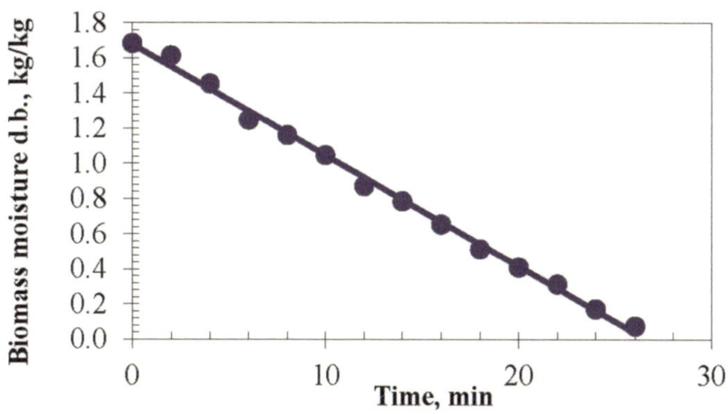


**Figure 9.**
*Drying curve for biomass particles in an agitated fluidized bed;* $R^2$ = *0.996.*

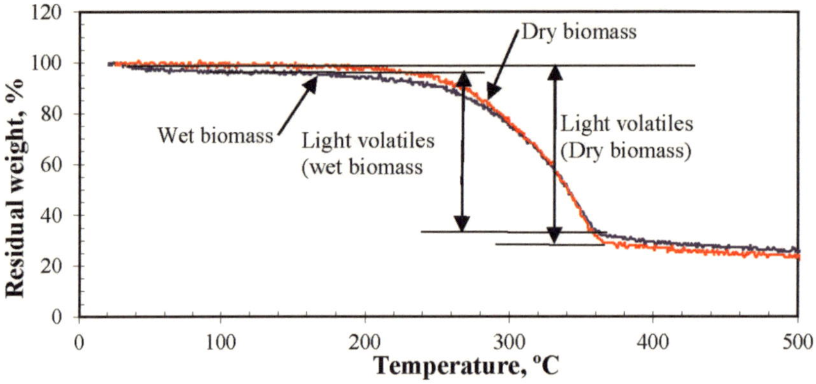


**Figure 10.**
*Thermogravimetric (TG) of wet and dry biomass in an inert atmosphere depending on the temperature.*

It can be seen in **Figure 10**, how the absolute loss of volatiles in the wet biomass is less than in the dry one, for having introduced less dry mass into the sample, since part of it was water (moisture). If the relative value were calculated, it would give the same proportion.

**Figure 11** shows two curves obtained for the test carried out in an oxidizing atmosphere, showing good reproducibility, which gives a good degree of reliability in the results obtained. The weight variations between the original test and the replica were 5.3%.

**Table 4** contains a summary of the relevant parameters of the thermogravimetric analysis. The maximum rate of devolatilization and combustion is evaluated according to the following expression:

$$U_{\max} = -\frac{1}{m_{isc}}\left(\frac{dm}{dt}\right)_{\max} \tag{28}$$

where $m_{isc}$ corresponds to the initial mass of loaded biomass, on an ash-free basis and $(\mathrm{d}m/\mathrm{d}t)_{\text{máx}}$ represents the maximum slope of the TG curve (mass versus time).

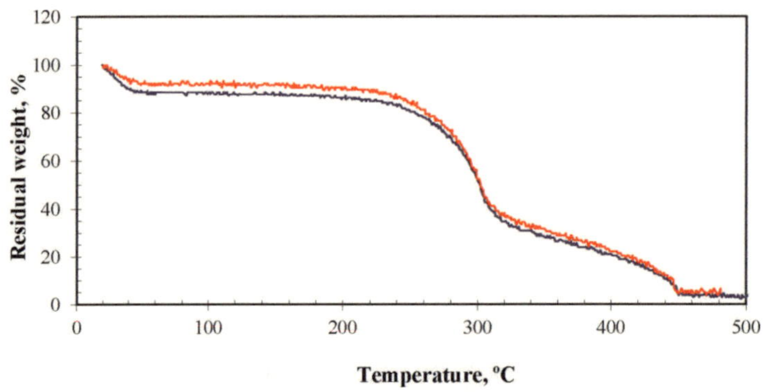


**Figure 11.**
*Thermogravimetric (TG) of biomass in an oxidant atmosphere and its replica.*

| Parameter | Inert atmosphere test | Air atmosphere test |
|---|---|---|
| Initial mass loaded (mg) | 5.95 | 6.29 |
| Ash content (%) | 3.90 | 3.90 |
| Initial mass without ash (mg) | 5.72 | 6.04 |
| Weight loss (%) and range 1 of T (°C) | 4.03 (21–110°C) | 12.24 (21–114°C) |
| Weight loss (%) and range 2 of T (°C) | 64.04 (155–364°C) | 55.96 (187–330°C) |
| Weight loss (%) and range 3 of T (°C) | 12.77 (364–800°C) | 27.90 (330–450°C) |
| Temperature of $U_{máx}$ (1) (°C) | 322 | 295 |
| Maximum rate (1) (g/g h) | 3.35 | 3.26 |
| Temperature of $U_{max}$ (2) (°C) | 574 | 437 |
| Maximum rate (2) (g/g h) | 0.18 | 1.13 |

**Table 4.**
*Thermogravimetric parameters derived from TG and DTG curves of forest biomass.*

### *4.2.2 Convective heat transfer coefficient*

The values obtained for the heat transfer coefficients are shown in **Table 5**, for each of the tests defined in **Table 1**. When using Eq. (15) in the calculation of the convective heat transport coefficient, the drying rate d*w*/d*t* shown in **Table 5** was introduced, which in turn is obtained from the experimental drying curve of each test (**Figure 9**). The value of $R^2$ indicated in the table corresponds to the correlation coefficient obtained in the fit of the straight line to the experimental data of moisture content of the solids versus time.

On the other hand, once the values of the heat transfer coefficient were known, in each test the Nusselt number was calculated using the first equality of Eq. (3) and then the correlation between the Nusselt number and the Reynolds number was obtained, which is proposed in Eq. (29). It has been verified that the rotation speed (*N*) does not

| $d_p$ | $D_p$ | $Re_p$ | -d*w*/d*t* | $R^2$ | $h_{gp\ exp}$ | $h_{gp\ anal}$ | $k_{gp\ exp}$ | $k_{gp\ anal}$ |
|---|---|---|---|---|---|---|---|---|
| (mm) | (mm) | | (kg kg$^{-1}$ min$^{-1}$) | | (W m$^{-2}$ K$^{-1}$) | (W m$^{-2}$ K$^{-1}$) | (m s$^{-1}$) | (m s$^{-1}$) |
| 0.89 | 2.44 | 102 | 0.067 | 0.99 | 15.4 | 13.0 | 0.010 | 0.009 |
| 0.89 | 2.44 | 103 | 0.0565 | 0.994 | 13.0 | 13.1 | 0.006 | 0.009 |
| 1.44 | 2.70 | 124 | 0.0553 | 0.986 | 13.2 | 15.0 | 0.008 | 0.011 |
| 1.44 | 2.70 | 124 | 0.0586 | 0.997 | 14.5 | 15.0 | 0.014 | 0.011 |
| 1.85 | 2.84 | 136 | 0.064 | 0.985 | 15.7 | 16.1 | 0.015 | 0.012 |
| 1.85 | 2.84 | 134 | 0.0631 | 0.996 | 14.9 | 15.8 | 0.015 | 0.012 |
| 2.18 | 3.13 | 148 | 0.0773 | 0.973 | 18.2 | 16.3 | 0.016 | 0.012 |
| 2.18 | 3.13 | 150 | 0.0645 | 0.996 | 14.4 | 16.5 | 0.012 | 0.013 |
| 3.56 | 4.30 | 256 | 0.0957 | 0.985 | 23.2 | 23.9 | 0.020 | 0.019 |
| 3.56 | 4.30 | 257 | 0.0992 | 0.99 | 25.7 | 24.0 | 0.015 | 0.019 |

**Table 5.**
*Convective heat and mass transfer coefficients.*

influence the convective heat transfer coefficient, since the variations found in the $h_{gp}$ coefficient as a function of $N$ are random as shown in **Table 5**, which is consistent with previously reported results [15].

$$Nu_{gp} = 0.003 Re_p^{1.28}; 100 < Re_p < 250; R^2 = 0.95 \tag{29}$$

or by rearranging terms as,

$$h_{gp} = 0.003 \frac{k_g}{D_p} \left( \frac{\rho_g U D_p}{\mu_g} \right)^{1.28} \tag{30}$$

**Table 5** shows the values found for the $h_{gp}$ coefficient, which fluctuate between 13 and 25.7 W $m^{-2}$ $K^{-1}$; these values are in a range similar to that of 6 and 23 W $m^{-2}$ $K^{-1}$ reported by Botterill [19]. In another investigation [20], results on the drying of forest biomass particles are reported, with $h_{gc}$ values in the range between 10 and 60 W $m^{-2}$ $K^{-1}$, to a dryer with superheated steam and operating temperatures up to 513 K. In a subsequent study by Salve et al. [21], higher $h_{gp}$ coefficient values (80–220 W $m^{-2}$ $K^{-1}$) were found, obtained with superficial velocities as well as elevated (3–11 m $s^{-1}$) and using sand in the bed.

On the other hand, **Table 6** shows a comparison between the values of the Nusselt number, obtained through the experimental procedure of this research, and those predicted by Eq. (29) with $R^2$ equal to 0.95. Furthermore, the experimental results have been compared with values predicted by Eq. (31) reported by Reyes and Alvarez [22] and obtained for Reynolds numbers in the range between 33 and 150.

$$Nu_{gp} = 0.00116\ Re_p^{1.52} \tag{31}$$

The comparison is also made with Eq. (32) from Zabrodsky and Eq. (33) by Lykov, respectively, both reported in Ciesielczyk [15].

| Re | $Nu_{gp}$ *(Experimental)* | Eq. (29) | Reyes and Alvarez, Eq. (31) | Zabrodsky, Eq. (32) | Lykov, Eq. (33) | Rao and Sen Gupta, Eq. (34) |
|---|---|---|---|---|---|---|
| 102 | 1.33 | 1.12 | 1.31 | 1.67 | 0.42 | 0.13 |
| 103 | 1.1 | 1.13 | 1.33 | 1.69 | 0.43 | 0.13 |
| 124 | 1.26 | 1.43 | 1.76 | 2.22 | 0.50 | 0.17 |
| 124 | 1.39 | 1.43 | 1.76 | 2.22 | 0.50 | 0.17 |
| 136 | 1.57 | 1.61 | 2.03 | 2.54 | 0.54 | 0.20 |
| 134 | 1.5 | 1.58 | 1.98 | 2.49 | 0.53 | 0.20 |
| 148 | 2 | 1.80 | 2.31 | 2.87 | 0.58 | 0.23 |
| 150 | 1.6 | 1.83 | 2.36 | 2.93 | 0.59 | 0.24 |
| 256 | 3.52 | 3.63 | 5.31 | 6.40 | 0.92 | 0.56 |
| 257 | 3.9 | 3.65 | 5.34 | 6.43 | 0.92 | 0.56 |

**Table 6.**
*Nusselt number: comparison of experimental values with the proposed equation and correlations from other authors.*

$$Nu_{gp} = 0.00195\ Re_p^{1.46} \tag{32}$$

$$Nu_{gp} = 0.0087\ Re_p^{0.84} \tag{33}$$

Finally, the results are compared with Eq. (34) from Rao and Sen Gupta and reported by Vyas and Nageshwar [23] with the Reynolds numbers in the range between 7 and 20.

$$Nu_{gp} = 0.000075\ Re_p^{1.61} \tag{34}$$

Of these correlations, those that are closest to Eq. (29) and the experimental values of this study are those of Reyes and Alvarez and that of Zabrodsky, Eqs. (31) and (32), respectively. The Rao and Sen Gupta equation is the one with the most deviations and is attributed to the fact that it is obtained for a very low range of the Reynolds number.

### *4.2.3 Convective mass transfer coefficient*

The experimental values of the convective coefficient of mass transfer under the conditions already established are shown in **Table 7**. As in the phenomenon of heat transfer, it is verified that the rotation speed of the mechanical stirrer does not affect the mass transfer.

In Eq. (35) the correlation obtained by adjustment between the Sherwood number and the Reynolds number is shown.

$$Sh_{gp} = 1.6x10^{-3} Re_p{}^{1.38}; \quad 100 < Re_p < 250; \quad R^2 = 0.75 \tag{35}$$

or:

$$k_{gp} = 1.6x10^{-3} \frac{D_v}{D_p} \left( \frac{\rho_g U D_p}{\mu_g} \right)^{1.38} \tag{36}$$

| $Re_p$ | $Sh_{gp}$ (experimental) | Eq. (31) | Froessling, Eq. (6) | Ranz and Marshall, Eq. (7) | Richardson and Szekely, Eq. (9) |
|---|---|---|---|---|---|
| 102 | 0.98 | 0.95 | 7.1 | 17.4 | 20.3 |
| 103 | 0.62 | 0.96 | 7.1 | 17.4 | 20.4 |
| 124 | 0.83 | 1.24 | 7.6 | 18.9 | 22.4 |
| 124 | 1.56 | 1.24 | 7.6 | 18.9 | 22.4 |
| 136 | 1.69 | 1.41 | 7.9 | 19.7 | 23.4 |
| 134 | 1.66 | 1.38 | 7.9 | 19.6 | 23.3 |
| 148 | 1.99 | 1.58 | 8.2 | 20.5 | 24.5 |
| 150 | 1.53 | 1.61 | 8.2 | 20.6 | 24.6 |
| 256 | 3.52 | 3.37 | 10.1 | 26.3 | 32.2 |
| 257 | 2.54 | 3.39 | 10.1 | 26.4 | 32.2 |

**Table 7.**
*Sherwood number: comparison of experimental values with the proposed equation and correlations from other authors.*

**Table** 7 shows a comparison between the values of the Sherwood number, obtained in this research, and those predicted by Eq. (35) with a correlation $R^2$ equal to 0.75. When comparing the experimental results with Eqs. (6), (7), and (9), unlike what happens with the heat transfer mechanism, in mass transport more significant differences are seen between our results and the Sherwood number values predicted by correlations obtained by other authors. Considering also that there are great differences between correlations already reported in the literature, it can be concluded that it is a more complex phenomenon to analyze and study experimentally. This is also verified in our research on mass transfer by observing that the value of $R^2$ from Eq. (35) is lower relative to the best correlation coefficient $R^2$ obtained in Eq. (29) for heat transfer.

The above implies that the information available in the specialized literature is not necessarily sufficient to predict heat and mass transfer coefficients, formulate mathematical modeling, and develop the thermal design of particulate forest biomass drying processes.

In the analysis of the discrepancies between experimental results with the values that predict equations reported in the literature, the characteristics of the solids of each system analyzed must be taken in consideration. It is also important to keep in mind the experimental procedures to determine the differences in temperatures and vapor concentrations between the fluidizing gas and the particles and the type of flow pattern in the fluidized bed (plug flow or complete mixing).

In this particular case, the system studied corresponds to a fluidized bed of coarse type D particles of the Geldart classification [24], which forces them to be fluidized with higher fluidization velocities and Reynolds numbers, which in this research ranged between 102 and 257.

Furthermore, Geldart type D biomass particles are solids with a high tendency to agglomerate as a result of cohesion forces between them, which are generated by surface water bridges. This is confirmed in this research with the lower values obtained for the specific surface area of wet particles $S_p$, compared to the values obtained when they are in a condition of low humidity $S_{p,0}$ (**Table 2**).

Regarding the flow pattern of the fluidized bed, it can be said that a fluidized bed of biomass particles, as a result of the combination of the air flow and the effect of the mechanical agitator, it could be considered as a system with complete mixing. Thus, according to the experimental results obtained on the *QF* fluidization quality index and temperature profiles inside the bed.

## 5. Conclusions

In an experimental equipment for drying particles in a fluidized bed, this research carries out experimental studies on the fluodynamic behavior of wet forest biomass particles suspended with a flow of hot gas in a fluidized bed with a mechanical stirrer. In addition, the biomass drying process and the heat and mass transfer mechanisms at the gas-particle interface are analyzed.

It is shown that the agitation has an important effect on the biomass particle aerodynamics if the particles have high moisture contents. The effect of the agitator is not relevant when fluidizing low moisture particles.

A methodology for the calculation of the biomass particle surface area is proposed based on Ergun equation. For the prediction of the $U_{mf}$ velocity, a new correlation is also proposed, which allows fitting the experimental values with ±15% deviations in the predictions of $U_{mf}$ for dry as well as wet particles.

On the other hand, the convective heat transfer coefficient between the gas and the solids was experimentally determined, varying between 13 and 25.7 W $m^{-2}$ $K^{-1}$. Based on a mass balance and on the experimental determination of the drying rates, the mass transfer coefficient was obtained, which varied between $6 \times 10^{-3}$ and $20 \times 10^{-3}$ m $s^{-1}$.

Thus, a correlation between the Nusselt number and the Reynolds number is proposed for the calculation of the heat transfer coefficient and a correlation between the Sherwood number and the Reynolds number for calculations of the mass transfer coefficient.

## Acknowledgements

This research was financially supported by the Universidad Austral de Chile and the Universidad de Valladolid in Spain.

## Nomenclature

| | |
|---|---|
| $A$ | cross section of the empty bed, $m^2$ |
| $A_p$ | superficial area of an arbitrary particle, $m^2$ |
| $c_v$ | moisture concentration in the air, kg $m^{-3}$ |
| $c_{v,i}$ | moisture concentration in the air at the entrance of the bed, kg $m^{-3}$ |
| $c_{v,o}$ | moisture concentration in the air at the exit of the bed, kg $m^{-3}$ |
| $c_{v,s}$ | moisture concentration in the air on the particle surface, kg $m^{-3}$ |
| $c_{v,\infty}$ | moisture concentration in the air outside the boundary layer, kg $m^{-3}$ |
| $C_g$ | gas heat capacity, J $kg^{-1}$ $K^{-1}$ |
| $d_p$ | particle size according to the sieving method, m |
| $D_p$ | diameter of the equivalent-volume sphere, m |
| $D_v$ | diffusivity of the vapor in the air, $m^2$ $s^{-1}$ |
| $D_0$ | reference particle diameter, m |
| $g$ | acceleration due to gravity, m $s^{-2}$ |
| $G$ | fluid mass velocity, kg $s^{-1}$ $m^{-2}$ |
| $h_{fg}$ | water heat of vaporization, J $kg^{-1}$ |
| $h_{gp}$ | heat transfer coefficient at the gas-particle interface, W $m^{-2}$ $K^{-1}$ |
| $k_g$ | thermal conductivity of the gas, W $m^{-1}$ $K^{-1}$ |
| $k_{gp}$ | mass transfer coefficient at the gas-particle interface, m $s^{-1}$ |
| $l$ | height in the fluidized bed, from the distributor, m |
| $L$ | bed height, m |
| $\dot{m}_{da}$ | dry air mass flow rate in the bed, kg $s^{-1}$. |
| $\dot{m}_v$ | water vapor flow rate between the solid and the gas, kg $s^{-1}$. |
| $N$ | agitation speed, rev $s^{-1}$ |
| $p$ | pressure, Pa |
| $S$ | particle surface area per unit bed volume, $m^2$ $m^{-3}$ |
| $S_{p,0}$ | particle surface area per unit volume of dry solids, $m^2$ $m^{-3}$ |
| $S_p$ | particle surface area per unit volume of solids, $m^2$ $m^{-3}$ |
| $t$ | time, s |
| $T$ | temperature, °C |
| $T_g$ | gas temperature, °C |
| $T_{g,i}$ | inlet gas temperature in the bed, °C |

| | |
|---|---|
| $T_{g,o}$ | outlet gas temperature in the bed, °C |
| $T_{g,l}$ | temperature of the gas in the bed at a height $l$, °C |
| $T_{p,s}$ | particle surface temperature, °C |
| $T_{wb}$ | wet bulb temperature, °C |
| $U$ | superficial velocity, m $s^{-1}$ |
| $U_{mf}$ | minimum fluidization velocity, m $s^{-1}$ |
| $v_a$ | specific volume of moist air, $m^3$ $kg^{-1}$ |
| $V_p$ | particle volume, $m^3$ |
| $w$ | moisture content in the biomass d.b., kg $kg^{-1}$ |
| $w_{ah}$ | humidity ratio of moist air d.b., kg $kg^{-1}$ |
| $W$ | particles weight in the bed, N |
| $\Delta c_{ml}$ | logarithmic-mean difference of concentration of vapor in moist air, kg $m^{-3}$ |
| $\Delta p$ | bed pressure drop, Pa |
| $\Delta T_{ml}$ | logarithmic-mean temperature difference, K |
| $\varepsilon$ | porosity of bed, $m^3$ $m^{-3}$ |
| $\mu_g$ | absolute viscosity of the gas, N s $m^{-2}$ |
| $\rho_b$ | bed density, kg $m^{-3}$ |
| $\rho_g$ | gas density, kg $m^{-3}$ |
| $\rho_p$ | biomass particle density, kg $m^{-3}$ |
| $\rho_{p,0}$ | dry particles density, kg $m^{-3}$ |
| $\phi$ | sphericity |
| $Ga$ | Galileo number, $D_p^3 \rho_g^2 g / \mu_g^2$ |
| $M_\rho$ | density ratio, $\rho_p/\rho_g$ - 1 |
| $Nu_{gp}$ | Nusselt number, $h_{gp} D_p/k_g$ |
| $Pr_g$ | gas Prandtl number, $C_g \mu_g/k_g$ |
| $Re_p$ | Reynolds number, $\rho_g U D_p/\mu_g$ |
| $Sc_g$ | gas Schmidt number, $\mu_g/\rho_g D_v$ |
| $Shgp$ | Sherwood number, $kgp\ Dp/Dv$ |

## Author details

Rogelio Moreno-Muñoz[1]*, Gregorio Antolín-Giraldo[2] and Alejandro Reyes-Salinas[3]

1 Instituto de Materiales y Procesos Termomecánicos, Universidad Austral de Chile, Chile

2 Departamento de Ingeniería Química y Tecnología del Medio Ambiente, Universidad de Valladolid, Spain

3 Departamento de Ingeniería Química, Universidad de Santiago de Chile, Chile

*Address all correspondence to: rmoreno@uach.cl

## References

[1] Corella J, Aznar MP, Delgado J, Aldea E. Steam gasification of cellulosic wastes in a fluidized bed with downstream vessels. Industrial & Engineering Chemistry Research. 1991; **30**:2252-2262

[2] Herguido J, Corella J, García JE, Artal G, Rodríguez JL. Gasificación de residuos sólidos orgánicos en sistema circulante multisólido [Gasification of organic solids wastes in circulanting multisolid sistem.]. In: *Actas I Congreso Internacional de Química de la Anque,* Islas Canarias. Vol. 3. 1990. pp. 485-491

[3] Pang A. Mathematical modelling of MDF fiber drying: Drying optimization. Drying Technology. 2000;**18**(7): 1433-1448

[4] Zabaniotou AA. Simulation of forestry biomass drying in a rotary dryer. Drying Technology. 2000;**18**(7): 1415-1431

[5] Moreno R, Antolín G, Reyes A. Thermal behavior of forest biomass drying in a mechanically agitated fluidised bed. Latin American Applied Research. 2007;**37**:105-113

[6] Ergun S. Fluid flow through packed columns. Chemical Engineering Progress. 1952;**48**(2):89-94

[7] Ranz WE, Marshall WR Jr. Evaporation from drops. Part I. Chemical Engineering Progress. 1952; **48**(3):141-146

[8] Kunii DO, Levenspiel O. Fluidization Engineering. New York: John Wiley & Sons, Inc.; 1969

[9] Froessling N. Ueber die verdunstung fallender tropfen. Gerlands Beitrage zur Geophysik. 1938;**52**:170-216

[10] Richardson JF, Szekely J. Mass transfer in a fluidized bed. Transactions of the Institution of Chemical Engineers. 1961;**39**:212-222

[11] Moreno R, Antolín G, Reyes A. Mass transfer during forest biomass particles in an agitated fluidised bed. Biosystems Engineering. 2020;**198**:163-171

[12] Vanecek V, Markvart M, Drbohlav R. Fluidized Bed Drying. London: Ed. Leonard Hill; 1966

[13] Temple SJ, Van Boxtel AJB. Modelling of fluidized-bed drying of black tea. Journal of Agricultural Engineering Research. 1999;**74**:203-212

[14] Zabrodsky SS. Hydrodynamics and Heat Transfer in Fluidized Beds. Cambridge, Massachusetts and London: The M.I.T. Press; 1966

[15] Ciesielczyk W. Analogy of heat and mass transfer during constant rate period in fluidized bed drying. Drying Technology. 1996;**14**(2):217-230

[16] Moreno R, Antolín G, Reyes A. Heat transfer during forest biomass particles in an agitated fluidised bed. Biosystems Engineering. 2016;**151**:65-71

[17] Watano S, Yeh N, Miyanami K. Heat transfer and the mechanism of drying in agitation fluidized bed. Chemical & Pharmaceutical Bulletin. 1999;**47**(6): 843-846

[18] Tannous K, Hemati M, Laguerie C. Fluidisation de grosses particules. In: Fluidización y Sistemas Fluido/ Partícula II. Zaragoza: Universidad de Zaragoza; 1992. pp. 41-48

[19] Botterill JSM. Fluid-Bed Heat Transfer. London: Academic Press; 1975

[20] Renström R, Berghel J. Drying of sawdust in an atmospheric pressure spouted bed steam dryer. Drying Technology. 2002;**20**(2):449-464

[21] Salve AM, Pande SS, Khan JS. Effect of velocity and particle size on the coefficient of heat transfer in fluidized bed exchanger. International Journal of Engineering Research and General Science. 2014;**2**(2):230-234

[22] Reyes A, Alvarez P. Determinación experimental de coeficiente de transferencia de calor en lechos fluidizados. In: Anais III Encontro Nacional de Ciencias Térmicas. Itapema SC: Universidade Federal de Santa Catarina; 1990. pp. 721-724

[23] Vyas RP, Nageshwar GD. Generalized correlations for steady and transient states gas to particle heat transfer in fluidized bed. IE (I) Journal-CH. 1999;**79**(March):62-67

[24] Moreno R, Antolin G, Reyes A, Alvarez P. Drying characteristics of forest biomass particles of Pinus radiata. Biosystems Engineering. 2004;**88**(1): 105-115

Chapter 3

# Improving the Efficiency of Rice Drying: Impact of Operational Variables on the Drying Rate and Quality of a South American Variety

*Laura Garcia-Llobodanin and Alejandra Billiris*


## Abstract

A key challenge for the rice industry during harvest is to improve the efficiency of the drying process, which involves increasing the drying rate and the head rice yield (HRY). In the present chapter, the main variables affecting the efficiency of rice drying were discussed. Then, the impact of the drying air conditions on the drying efficiency of a long-grain South American rice variety at different rice moisture contents (MC) was studied using a thin-layer lab-scale dryer. Drying at each air condition was modeled using Page's equation. The drying rate increased as the air conditions became more extreme (higher temperature or lower relative humidity). The effect on the HRY depended on the rice MC. Therefore, a two-stage drying program was proposed using different drying air temperatures depending on rice MC. These results were applied to create a drying program for a long-grain South American variety dried in a cross-flow commercial dryer. The two programs tested increased the drying rate and one of them also increased the HRY, compared to drying at the industry operational conditions. Implementing this program would improve the efficiency of the drying process, increasing the reception capacity and the profitability of the rice obtained.


**Keywords:** rice drying, drying rate, drying efficiency, rice quality, drying optimization

## 1. Introduction

Rice is a staple crop that feeds almost half of the world population [1]. It is usually harvested at a moisture content (MC) of 16 to 22% and needs to be dried to a MC of 13% or lower for safe storage [2]. Considering that rice must be dried immediately after harvest to prevent spoilage, drying becomes a critical process during the harvest season.

Increasing the drying rate is relevant because it allows a larger amount of rice to be dried in a given period of time, relieving the frequently occurring bottleneck generated when rows of producers' trucks are waiting to deliver their wet rice at the industrial plant.

During drying, moisture is removed from the surface of the rice grain. Initially, there is enough water available at the surface, making the drying rate rapid. After a short period of time, drying is limited by water diffusion from the interior to the surface of the kernel. This generates an increasing moisture gradient inside the kernel, with the center having a higher MC than the surface [1]. MC gradients are present during the entire drying process, and they depend on the drying rate, which is determined by the rice grain MC and the drying air conditions (temperature, relative humidity, and flow rate) [3]. The surface of the kernel tends to the equilibrium moisture content (EMC), which is given by the drying air temperature (T) and relative humidity (RH) [4, 5]. A higher T or lower RH decreases the EMC, increasing the drying rate and the MC gradient.

The MC gradient generates tension at the surface of the grain, where the cells tend to shrink as moisture is lost, compressing the center [6]. This is associated to the formation of fissures and cracking during drying. Fissured kernels are more prone to breaking during milling. Therefore, the head rice yield (HRY), defined as the mass percentage of rough rice that remains as head rice (kernels that are at least three-fourth of its original length) after milling, tends to decrease with the presence of fissured kernels. In addition, fissured kernels affect the functional properties of milled rice, and thus the sensory quality [5].

Glass transition temperature (Tg) also plays an important role in fissures formation. Tg is the temperature range of transition of the amorphous regions of starch from a glassy into a rubbery state. Starch is the main component of rice and is formed by amylose and amylopectin chains. Glass transition occurs at the branching points of amylopectin [7]. The glassy state has a low expansion coefficient, specific volume, specific heat, and diffusivity but high viscosity and modulus of elasticity. Contrarily, the rubbery state has higher expansion coefficient, specific heat, specific volume, and diffusivity [5, 8]. The Tg increases as the grain MC decreases. During drying, the MC at the surface of the grain is lower than that at the center. This could cause the center of the grain to be in the rubbery state, while the surface is in the glassy state. The differences in the properties of the two states increase the tensions generated by the MC gradient and play an important role in terms of kernel fissuring potential [8].

To prevent this, at least in part, a process called tempering is used between drying passes. During tempering, rice is held in bins for a certain period of time. The purpose is to allow the MC gradients generated during drying to subside, reducing the tensions inside the kernels and therefore preventing kernels' fissuring [5, 9].

The drying rate is also affected by grain composition and geometry. Therefore, different varieties may respond differently to the same drying air conditions [1, 10].

As a result of this, the need arises to find suitable drying programs for each variety, reducing the drying time while minimizing fissures formation. A compromise should be made between the drying rate and the MC gradient generated during drying, which could lead to fissures formation, especially when two states (glassy and rubbery) coexist within the same kernel.

Several authors studied the impact of the drying air conditions and Tg on the drying rate and HRY [2, 5, 11]. Most of the studies were conducted using long-grain rice laid out in a thin layer, to ensure that all the rice is subjected to the same conditions.

As previously exposed, rice variety also plays an important role in relation to drying. Long-grain and medium-grain rice showed different behavior during drying

[10]. This was attributed to differences in the kernels' dimensions. Medium-grain kernels were thicker, so moisture had to travel a longer path in its migration to the surface (compared to long-grain kernels).

Different drying methods, including experiments in commercial dryers, were also studied by some researchers. In Ref. [12], continuous drying of rough rice was compared with intermittent drying, while [13] investigated rough rice drying in fixed and fluidized bed dryers. Natural drying (shade and sun drying) was compared with heated air drying in [14] using different drying methods in a commercial and a lab scale.

The research published so far is mostly on varieties developed in the United States or Asia. There is very little literature on South American varieties, which have their own characteristics given by climatic conditions, cultivation practices, and genetics.

The present chapter introduces a review on the main variables affecting the drying efficiency, understanding the drying efficiency as the combination of drying rate and HRY. Then, the impact of the operating conditions on the efficiency of rice drying was studied for a South American variety using a thin-layer lab-scale dryer. Finally, an industrial application of the previous results is shown for a South American variety dried in a commercial cross-flow dryer.

## 2. Variables affecting the efficiency of rice drying

This section presents a review of the main variables affecting the drying efficiency, understanding the drying efficiency as the combination of drying rate and HRY.

### 2.1 Drying air conditions

The drying air conditions have an impact on the drying rate and the HRY. Severe drying conditions (high drying air temperatures, flow rate, and/or low RH) increase the MC gradient, increasing the drying rate but decreasing the HRY.

In Ref. [10], the authors study the HRY of rice dried at different drying conditions. At the milder condition (higher EMC), the HRY was not affected during the entire drying process. However, as the drying conditions became more severe (lower EMC), the HRY decreased during drying. The more severe the drying condition, the sooner the HRY begins to decrease once drying starts. This could probably be attributed to the formation of more pronounced MC gradients.

An increase in the drying air temperature increases the drying rate, even for the same EMC [2]. The HRY is also susceptible to air temperature, even with equal EMC conditions. This is probably because higher drying rates cause more pronounced MC gradients inside the kernel, increasing tensions and favoring fissures formation. The HRY reduction is even more pronounced when drying above Tg, probably due to the coexistence of glassy and rubbery zones (surface and core, respectively) in the same rice kernel [15].

Regarding the drying air RH, when reduced (at a given temperature), the drying duration is shortened [16]. This could be related to an increase in the drying potential of the air.

### 2.2 Glass transition temperature

Tg is the temperature range corresponding to the transition of the amorphous regions of starch from a glassy to a rubbery state. It is a useful material descriptor due to its good correlation with structural and thermodynamic properties [17]. Tg depends on the composition of the rice grains, particularly starch. The amylose/amylopectin ratio plays an important role in the Tg. The higher the amylose content, the higher the Tg, which is associated to chain-chain interactions of linear chains of amylose that induce partial crystallinity [18]. On the contrary, high amylopectin starches show lower Tg. This was attributed to the formation of gel balls by the short-branched chains in amylopectin molecules. The gel balls require less energy to move than long linear chains, reducing the Tg. Water also reduces the Tg, acting as a plasticizer [7, 18]. Therefore, the Tg of a rice kernel depends on its MC. This means that at a certain temperature, a kernel could be in the glassy or in the rubbery state, depending on its MC.

The abrupt change in several properties of the material in the glass transition range can be used for its determination [17]. There are changes in two groups of properties: rheological and thermodynamic properties. Differential scanning calorimetry (DSC) is used to determine Tg based on changes in thermodynamical properties, such as heat capacity. Dynamic mechanical thermal analysis (DMTA) is another methodology used and is based on the change in rheological properties, such as the storage and loss moduli. Both methodologies showed good results and proved to be suitable for Tg determination [8, 19].

**Figure 1** shows the state diagram for three Uruguayan rice varieties (Uy1, Uy2, and Uy3), built using DSC [20]. A significative difference between the Tg of Uy1 and Uy3 in the MC range of 12 to 16% was found, proving that different varieties could have differences in the Tg. This was attributed to differences in the starch composition and in the kernels' dimensions of both varieties.

Tg plays an important role in the rice drying process. Drying at temperatures above the Tg increases the drying rate since thermal conductivity and mass diffusivity are higher in the rubbery state. However, the possible presence of glassy and rubbery regions within the same kernel (due to differences in the MC given by the MC

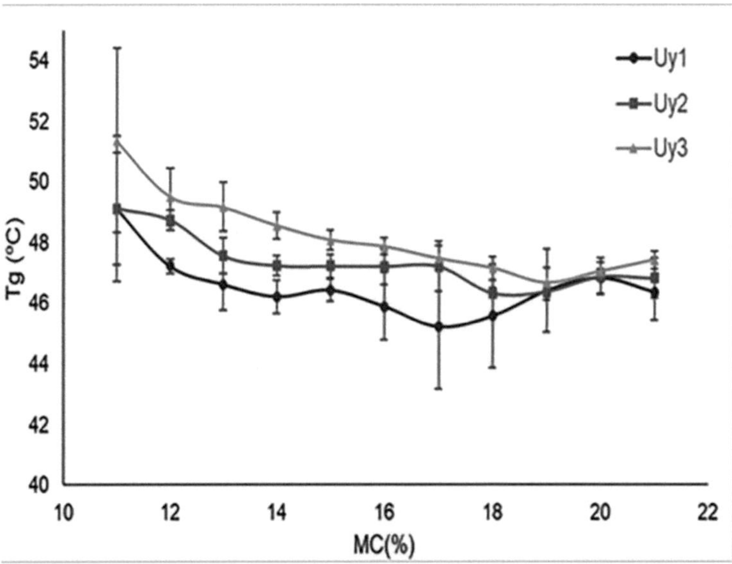


**Figure 1.**
*State diagram for three Uruguayan varieties (reprinted, with permission, from [16]). Tg, glass transition temperature; MC, moisture content.*

gradient formed) could increase the tensions inside the grain, favoring fissures formation and reduction of the HRY [4]. **Figure 2** represents this situation in a state diagram [15]. A rice kernel with initial MC of 19% is heated during drying to a temperature of 55°C, going from the glassy to the rubbery zone. As drying continues, the center (more humid) remains in the rubbery zone, while the surface tends to the EMC in the glassy zone. This increases the tensions that already exist inside the kernel due to the MC gradient. Increasing the air relative humidity (RH) increases the EMC, and consequently reduces the MC gradient. This enables a greater part of the kernel to remain in the rubbery state (when drying above the Tg), and therefore increases the HRY [4].

In Ref. [21], a glass transition mapping inside a rice kernel during drying was performed by modeling. It was found that when the drying air temperature is higher than the Tg, the outer layers of the kernel go from the glassy to the rubbery state due to a rapid temperature rise at the beginning of drying. Then, as MC decreases, a transition from the rubbery back to the glassy state could occur (as shown in **Figure 2**). Simulation suggests that fissures initiate more easily from the tensile zones, where the transitions from a rubbery to a glassy state occur, probably because of the stress concentration in the interface between the regions of expansion and contraction caused by the change of state.

### 2.3 Kernels dimensions and composition

The dimensions and composition of the grain kernels affect their behavior during drying. As the kernel thickness increases, the fissured kernel percentage also increases [22]. It was hypothesized that the thicker kernels experience a greater MC gradient than thinner kernels when exposed to the same drying condition. This leaves thicker kernels more susceptible to fissuring. Confirming these results, a study with Bengal (medium-grain), Cypress, and Kaybonnet (both long-grain) varieties dried under the

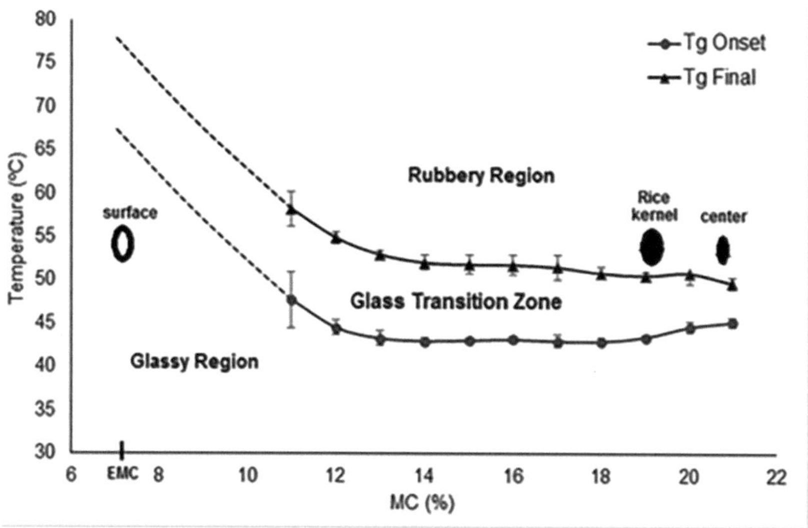


**Figure 2.**
*Representation of a rice kernel (whole grain, center, and surface) in a state diagram. MC, moisture content, EMC = equilibrium MC, Tg = glass transition temperature (reprinted, with permission, from reference [11]).*

same drying conditions shows that medium-grain Bengal has a faster and more pronounced HRY reduction compared to the other two varieties [10].

The chemical composition (amylose and protein content) and physicochemical properties of rice also impact the HRY [3]. Drying of both low and high amylose rice, at the same level of grain temperature, shows quality differences [14].

The drying rate is also affected by the rice grain variety, influenced by the kernel dimensions and/or composition [2].

## 3. Impact of operating conditions on the efficiency of rice drying

As seen in the previous section, there are numerous variables that affect the drying rate and the HRY, among which are the operating (drying air) conditions. The operating conditions needed to increase the drying rate are usually opposite to those needed to improve the HRY.

The aim of the present section is to study the impact of the operating conditions on the milling quality and drying duration of a South American long-grain rice variety (Uy2) at different stages of the drying process. To this purpose, a thin-layer lab-scale dryer was used to dry rice under different controlled drying air conditions. The drying process at each condition was modeled using Page's equation.

### 3.1 Materials and methods

#### *3.1.1 Rice sample*

Rice of the Uruguayan variety Uy2 was collected from a single producer in the south-east region of Uruguay. The MC was determined by gravimetry [23]. The harvest MC of the paddy lot was 20.5 ± 1.0%. The sample was homogenized and stored in a refrigerating chamber at 4.3 ± 1.8°C until use. Before each experiment, the amount needed was removed from the chamber and left in sealed bags at room temperature for at least two hours.

#### *3.1.2 Experimental procedure*

Drying runs of long-grain rice Uy2 at different MC levels and using different drying air conditions (temperature and relative humidity) were performed. The drying air velocity was set at 0.4 m/s for all runs and the air T and RH were constant during each run.

**Table 1** shows the experimental design. Runs were carried out at different temperatures, below and above the Tg, and two RH were tested at each T. The RH was chosen to have the same EMC values (7% and 10%) at all the air T levels, meaning that the grain surface MC was also the same (since the grain surface MC equals the EMC soon after the drying begins). The RHs for each combination of EMC and air T were calculated using the modified Chung-Pfost equation for long grains [24]. For some runs, the air conditions set could not be reached (due to limitations of the drying system). In those cases, the runs were set at the closest condition possible (see **Table 1**). At the air T of 47°C, a greater variation of the milling quality was observed between the two EMC conditions (compared to other air temperatures). Therefore, two extra runs at two different RHs (corresponding to EMC of 8% and 12%) were added.

| T (°C) | *RH* (%) | EMC (%) |
|---|---|---|
| 35 | 25 | 7.4 |
| 35 | 50 | 10.0 |
| 47 | 27 | 7.0 |
| 47 | 42 | 8.5 |
| 47 | 57 | 10.0 |
| 47 | 70 | 12.0 |
| 55 | 30 | 7.0 |
| 55 | 60 | 10.0 |
| 65 | 31 | 6.5 |
| 65 | 45 | 8.6 |

*T, Air temperature (°C); RH, Relative humidity (%); EMC, Equilibrium moisture content.*

**Table 1.**
*Experimental design.*

For each drying condition, rice was dried to a final MC of 17 ± 0.7%, 15 ± 0.7% and 13 ± 0.7%. This allowed the study of the impact of the drying conditions on the milling quality at each MC range.

The rice samples reached the drying air temperature within the first 2–3 minutes of run in all cases. Therefore, the grain temperature was considered equivalent to the drying air temperature.

### *3.1.3 Drying system*

Rice was dried in a laboratory drying equipment, especially designed and built for this purpose (Urumáquinas, Uruguay). It allowed controlling the drying air conditions including T, RH, and velocity with a precision of ±0.6°C, ±2.6%, and ±0.02 m/s, respectively. The equipment also monitored the weight loss and grain temperature of the sample during drying. **Figure 3** shows a schematic of the drying system used.

The ambient air entered the system with the aid of a blower, which controlled the air velocity. A condenser and a vapor injector regulated the air humidity and resistors regulated the air temperature. A sensor of T and RH together with a sensor of velocity

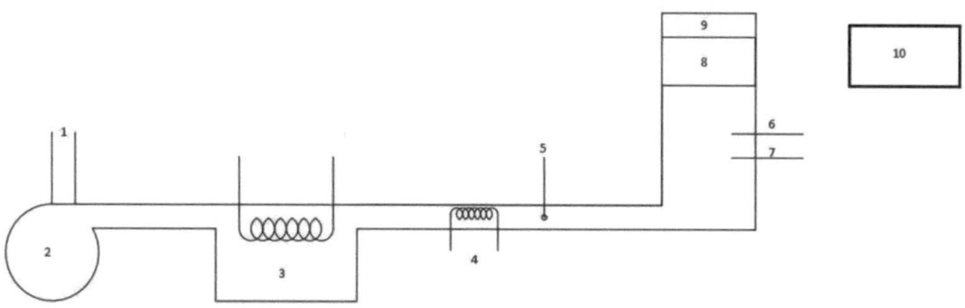


**Figure 3.**
*Schematic of the drying equipment. 1-air entrance, 2-blower; 3-condenser, 4-resistors, 5-vapor injector, 6-velocity sensor, 7-air temperature and relative humidity sensor, 8-drying chamber, 9-load cell, and 10-PLC.*

was installed just before the drying chamber. Air conditions were set and controlled with the aid of a PLC (Secoin, Uruguay).

The rice sample was disposed in a tray with a perforated bottom to allow the air circulation. A temperature sensor was introduced in the rice sample to monitor the grain temperature. The sample weight was measured with the aid of a load cell, and the grain MC was calculated at different times using the initial MC and the weight loss:

$$MCt = 100 \times \left(1 - \frac{IW}{Wt} \times \left(1 - \frac{IMC}{100}\right)\right) \quad (1)$$

where MCt is the MC at a time t, IW is the initial weight of the sample, Wt is the weight of the sample at a time t, and IMC is the initial MC of the sample expressed on a wet basis.

All parameters (drying air T, RH, velocity, sample temperature, and weight) were registered every five minutes along each drying run.

### *3.1.4 Rice drying*

Once the drying air reached the set condition, five hundred grams of paddy rice were put on the tray, arranged in a thin layer of one centimeter high, and introduced into the drying chamber.

A drying curve was built for each condition, leaving the rice to dry until no MC change was detected (at least ten consecutive measurements with grain MC differences among measurements lower than 0.5%). The drying curves were fitted to Page's Eq. (2):

$$\frac{MC - EMC}{IMC - EMC} = \exp(-k\ x\ t^{n}) \quad (2)$$

where MC is the moisture content at the drying duration t (h), EMC is the equilibrium moisture content, IMC is the initial moisture content, k is the drying rate constant ($h^{-1}$), and n is a dimensionless constant. MC, EMC, and IMC are expressed in a decimal dry basis.

Then, for each drying condition, rice was dried to a final MC of 17 $\pm$ 0.7%, 15 $\pm$ 0.7% and 13 $\pm$ 0.7%. The time needed to reach each final MC at each drying air condition was calculated using the fitted Page's equation.

After drying, samples were submitted to one-hour tempering at the correspondent drying air temperature. Based on preliminary experiments, this was the minimum tempering time needed to minimize kernels' breakage after drying [15]. After tempering, samples with a final MC of 17% and 15% were dried in a chamber (Alfa-Laval Gruppe, Germany) at 20.5°C and 60% RH until a final MC of 13 $\pm$ 0.5%. This gentle drying has a minimum impact on the grain quality, allowing to study the drying process at the MC range of interest. All experiments were performed in triplicate.

### *3.1.5 Milling quality*

After drying and tempering, the samples were kept at room temperature for at least 72 hours. Then, the head rice yield (HRY) was determined.

Before milling, each sample was cleaned with a grain cleaner (Grainman, USA). Then, 100 g of clean paddy rice was hulled using a paddy husker (THU35B, Satake,

Japan). The dehulled rice samples were milled with a laboratory rice polisher (TM05C, Satake, Japan) to a degree of milling (DOM) of 100 ± 3, measured with a milling meter (MM1D, Satake, Japan). After milling, the broken kernels were separated using a trieur (Satake, Japan) and quantified using an image analyzer (Image 5, Selgron, Brazil). The results were expressed as grams of head rice obtained from 100 g of rough rice.

The head rice yield reduction (HRYR) during drying was defined as:

$$HRYR = HRY_{final} - HRY_{initial} \tag{3}$$

where $HRY_{final}$ was the HRY after the drying/tempering process at each drying condition evaluated, and $HRY_{initial}$ represents the "maximum milling potential" of the rice lot. The HRYR is expressed in percentage points (pp), corresponding to the grams of milled head rice every 100 g of rough rice.

To determine the "maximum milling potential", four samples were gently dried in a chamber (Alfa-Laval Gruppe, Germany) at 20.5°C and 60% RH until a final MC of 13 ± 0.5%. This air condition produces minimum fissuring and thus, minimal quality loss [10, 11]. Therefore, the HRYR obtained represents the maximum milling quality that can be achieved for the rice lot used.

#### *3.1.6 Statistical analysis*

The drying curves were fitted to Page's equation using the software JMP 12.0.

The standard deviation was calculated for the Page's equation parameters and the HRYR. Analysis of variance (ANOVA) was used to compare the n constant of Page's equation at different drying conditions. In the case of significative difference ($p < 0.05$), the Tukey test was applied to determine which are the values that differ.

The mean squared error (MSE) was calculated for Page's equation at each drying condition.

### 3.2 Results and discussion

**Table 2** shows the EMC calculated using the modified Chung-Pfost equation, the parameters k, n, and EMC from the fitted Page's equations, the corresponding mean squared error (MSE), and the drying durations to reach a final MC of 13%, 15%, and 17% (calculated using the fitted Page's equations) for each drying air condition.

The modified Chung-Pfost equation has been extensively used to calculate the EMC of grains. In Ref. [24], five different equations were compared and their suitability for describing the EMC of rough rice of different varieties (long and medium grain) was evaluated in a wide range of T and RH. The Chung-Pfost equation was the best model for describing equilibrium data.

Mathematical modeling of rough rice drying has also been studied by several researchers. A diffusion model assuming that liquid diffusion is the only moisture transfer inside the rice kernels has been used by some authors [25, 26]. However, solving this type of modeling is quite complex. Therefore, researchers usually use empirical or semiempirical models to simulate rice drying [27]. In Ref. [28], ten different models for continuous and intermittent drying of thin-layer rough rice were compared. The authors found that the Midilli model showed the best results but another four of them, including Page's model, were also adequate in describing the experimental data.

| T (°C) | RH (%) | $EMC_{C\text{-}Pfost}$ (%) | $EMC_{Page}$ (%) | k ($h^{-1}$) | n | MSE | $t_{MC}$ = 13% (min) | $t_{MC}$ = 15% (min) | $t_{MC}$ = 17% (min) |
|---|---|---|---|---|---|---|---|---|---|
| 35 | 25 | 7.4 | 7.3 ± 0.4 | 0.47 ± 0.04 | $0.60 \pm 0.04^{A,B}$ | 0.0342 | 171 | 80 | 29 |
| 35 | 50 | 10.0 | 7.9 ± 0.4 | 0.35 ± 0.04 | $0.54 \pm 0.04^{B}$ | 0.1054 | 377 | 158 | 51 |
| 47 | 27 | 7.0 | 6.7 ± 0.4 | 0.69 ± 0.04 | $0.59 \pm 0.02^{A,B}$ | 0.0396 | 80 | 37 | 13 |
| 47 | 42 | 8.5 | 8.2 ± 0.4 | 0.69 ± 0.04 | $0.61 \pm 0.08^{A,B,C}$ | 0.0302 | 115 | 55 | 22 |
| 47 | 57 | 10.0 | 10.9 ± 0.2 | 0.54 ± 0.00 | $0.74 \pm 0.06^{D,E}$ | 0.0204 | 244 | 110 | 44 |
| 47 | 70 | 12.0 | 12.3 ± 0.6 | 0.61 ± 0.02 | $0.73 \pm 0.06^{D,E}$ | 0.0180 | 474 | 166 | 74 |
| 55 | 30 | 7.0 | 5.7 ± 2.0 | 0.76 ± 0.08 | $0.60 \pm 0.08^{A,B}$ | 0.0356 | 67 | 35 | 15 |
| 55 | 60 | 10.0 | 9.3 ± 0.4 | 0.84 ± 0.1 | $0.69 \pm 0.06^{C,D}$ | 0.0239 | 102 | 52 | 23 |
| 65 | 31 | 6.5 | 6.7 ± 0.2 | 1.16 ± 0.04 | $0.65 \pm 0.08^{A,C}$ | 0.0598 | 35 | 17 | 7 |
| 65 | 45 | 8.6 | 8.5 ± 0.4 | 1.17 ± 0.06 | $0.80 \pm 0.02^{E}$ | 0.0552 | 51 | 28 | 13 |

*Error: ± 2σ. Different letters within a column indicate significative difference. T, temperature; RH, relative humidity; EMC, equilibrium moisture content using the Chung-Pfost eq. (C-Pfost) and Page's equation (Page), k, kinetic constant; n, Page's equation constant; MSE, mean squared error; t, duration to reach the indicated moisture content (MC). Each experimental value is an average of three replicates.*

**Table 2.**
*EMC obtained from Chung-Pfost equation, Page's equation parameters (EMC, k, and n), MSE, and time needed to reach the final MC.*

In Ref. [27], the suitability of twelve empirical and semiempirical models in defining thin-layer drying behavior of long-grain rough rice was studied. They also found that Midilli's model showed the best results, in part due to its high number of coefficients (four). The fact that it is a simplified version of the diffusion equation could also contribute, proving that liquid diffusion is the dominant transport mechanism in rough rice. Nevertheless, they found that Page's model was also suitable and was the most accurate among the two-parameter models. In agreement with these results, Pereira et al. [26] found that Page's model was the best model, among the six models studied, for describing continuous and intermittent drying of rough rice. Based on these previous reports, Page's model could be considered a simple (only two parameters), suitable model for describing thin-layer rough rice drying.

In the present study, except for the air condition at T = 35°C and RH = 50%, the EMC calculated using the modified Chung-Pfost equation was in quite good agreement with those obtained from Page's equation. In addition, Page's equations presented low MSE values (see **Table 2**), confirming its suitability for representing thin-layer drying of the long-grain rice variety Uy2.

**Table 2** also shows that n values are significantly different among some of the runs ($p < 0.05$). The k values can only be compared among those runs with n not significantly different. In those cases, k increased as the drying air temperature increased. Consequently, at constant EMC, the time needed to reach a certain grain MC decreased as temperature increased, as shown in **Table 2**. In agreement with our results, Chen et al. [2] found that higher temperatures may cause higher k values, even when the EMC was the same. This behavior could be expected since higher drying temperatures are associated with higher drying rates [16]. This is probably due to higher moisture effective diffusivities at higher drying temperatures [27, 29].

In Ref. [9], the authors found that drying above the Tg significantly increased the drying rate compared to drying below the Tg. This was attributed to the higher diffusivity observed in the rubbery state (compared to the glassy state). As previously exposed, in our experiments the drying constant (k), when comparable, also increased as the drying air T increased. In fact, when the drying air T increased from 35°C (below Tg) to 47°C (above Tg) at a constant EMC of 7%, the value of k increased almost 50%. However, when the drying air T increased from 47 to 55°C (both above Tg), the value of k only increased 10%. Therefore, the sharper increase of k in the former situation could be attributed, at least in part, to the glass transition phenomenon.

**Table 3** presents the HRYR of the rice dried at different drying air conditions and to different final MC. For rice taken to 17% and 15% MC, drying was finished in the drying chamber under mild conditions (T = 20.5°C and 62% RH). It can be observed that rice could be dried to a MC of 15% using drying air at temperatures as high as 47°C and RH as low as 27% maintaining a low HRYR (under 5 pp). Drying air temperatures of 55°C or higher increased drastically the HRYR.

During the last stage of the drying process (15–13% MC), milder drying air conditions should be applied to maintain a low HRYR. At 35°C, the HRYR was low at both RH tested. However, at 47°C the RH should be 57% or higher to keep the HRYR low.

This research brings important information on how the drying air conditions affect the drying rate and HRYR of a South American long-grain rice variety. The results could be used to implement a drying program that improves both aspects, using a more severe drying air condition until a MC of 15% to increase the drying rate, and then softening the drying air conditions at MC below 15%, to avoid an increase in the HRYR.

| T (°C) | *RH* (%) | MC = 17% | MC = 15% | MC = 13% |
|---|---|---|---|---|
| 35 | 25 | 3.1 ± 1.4 | 2.0 ± 0.5 | 3.7 ± 0.9 |
| 35 | 50 | 2.7 ± 1.3 | 2.1 ± 1.3 | 1.5 ± 1.2 |
| 47 | 27 | 2.2 ± 0.7 | 2.5 ± 0.7 | 34.4 ± 3.3 |
| 47 | 42 | 4.1 ± 0.3 | 4.2 ± 0.8 | 8.6 ± 4.6 |
| 47 | 57 | 6.9 ± 0.3 | 3.8 ± 1.4 | 2.6 ± 0.6 |
| 47 | 70 | 3.5 ± 1.6 | 4.9 ± 1.8 | 4.4 ± 0.4 |
| 55 | 30 | 7.6 ± 1.9 | 10.9 ± 0.3 | 40.4 ± 1.7 |
| 55 | 60 | 7.4 ± 1.2 | 8.9 ± 1.2 | 26.5 ± 0.5 |
| 65 | 31 | 5.3 ± 0.9 | 9.4 ± 3.9 | 47.4 ± 1.6 |
| 65 | 45 | 5.1 ± 1.0 | 18.4 ± 3.4 | 56.0 ± 1.5 |

*HRYR is expressed in percentage points. Errors of HRYR correspond to two standard deviations (±2σ).*

**Table 3.**
*Head rice yield reduction (HRYR) of rice dried to different moisture contents (MC: 17%, 15%, 13%) at different drying air conditions (T, RH). Rice with MC 17% and 15% was taken to a final MC of 13% under mild drying conditions (T = 20.5 °C/62% RH).*

## 4. Industrial application: a case study

In the first two sections, a review of the main concepts and research works regarding the variables affecting rice drying were presented. In the third section, the impact of the operation variables on thin-layer drying of a South American rice variety was studied. Based on these results, in the present section two drying programs are proposed and tested to dry a South American variety in a commercial cross-flow dryer, with the aim of increasing the HRY and the drying rate. The objective of this section is to provide a practical application to the results obtained on a laboratory scale.

### 4.1 Materials and methods

#### *4.1.1 Commercial dryer*

Runs were performed in a cross-flow commercial dryer. The dryer has two sections: a drying chamber and a tempering zone. Rice enters the dryer and recirculates, passing the drying chamber and tempering zone in each cycle, until it reaches the final MC (approximately 13%). Contrarily to what occurs in the lab-scale dryer (see Section 3), in the commercial dryer, rice temperature never reaches the drying air temperature (due to the absence of a thin layer of rice and the passage through the tempering zone between the drying cycles). Therefore, the drying programs were based on controlling the rice grain temperature (not the drying air temperature).

The drying air was taken from the environment and brought to the desired drying temperature in an industrial oven. Therefore, its RH depends on the environmental conditions. Since the drying air temperature was much higher than the ambient, the RH of the drying air was low.

#### *4.1.2 Experimental design*

From the experiments in the lab-scale dryer (see Section 3), it could be concluded that at low RH (which is the case in the commercial dryer) and a grain MC of 15% or higher, the air temperature (which equals the grain temperature) should be below 55° C and above 47°C. This allows a high drying rate with low HRYR (below 5 pp). Higher drying air temperatures not only increase the drying rate but also increase the HRYR.

At grain MC below 15% and low RH, the air temperature should be below 47°C, being the HRYR very low at an air temperature of 35°C (no data is available between 35 and 47°C).

During the experiments, it was also observed that the Tg plays an important role, being critical when a pronounced MC gradient is present within the kernel. Therefore, drying at temperatures above Tg should be avoided at low grain MC.

Based on these considerations, two programs were proposed and tested by triplicate:

Program A:

- Grain MC > 18%: grain temperature = 48 ± 2°C
- Grain MC < 18%: grain temperature = 42 ± 2°C

Program B:

- Grain MC > 18%: grain temperature = 48 ± 2°C
- 15% < Grain MC < 18%: grain temperature = 42 ± 2°C
- Grain MC < 15%: T grain temperature = 40 ± 2°C

#### *4.1.3 Experimental procedure*

The rice used was a long-grain Uruguayan variety, with similar dimensions and composition than Uy2 (see Section 3). All runs (at the ordinary operating conditions and the programs) were performed with the same rice variety.

A representative sample of the rice entering the dryer in each run was collected using an automatic sampler. It collected a sample of approximately half kilogram every 2 minutes during the loading of the equipment. Once the loading was completed, the samples were mixed and homogenized properly. The same procedure was followed during the unloading, to obtain a representative sample at the exit of the dryer (once the drying run was finished). The rice MC was determined by gravimetry [23], and the HRYR was determined as described in Section 3.1.5, being $HRY_{final}$ the HRY of the sample collected at the exit of the dryer and $HRY_{initial}$ the HRY of the sample collected at the entrance of the dryer and gently dried in the chamber until a final MC of 13 ± 0.5%.

The average drying rate of each run was defined as the average MC (on a dry basis) removed per hour and was calculated as:

$$Drying\ rate = \frac{MCfinal, d.b. - MCinicial, d.b.}{time} \tag{4}$$

where MCfinal,d.b. is the MC of rice at the end of the run on a dry basis (%MC), MCinitial,d.b. is the MC of rice at the beginning of the run on a dry basis (%MC), and time is the drying duration of the run (hours).

Thermal properties of all the rice samples were measured with a differential scanning calorimeter (TA instruments, DSC Q2000) as described in [30]. The data obtained comprise the onset temperature, peak temperature, conclusion temperature, and crystal melting enthalpy ($\Delta H$).

Pasting properties were measured with a rapid visco analyzer (Perten Instruments, RVA 4500) following AACCI Approved Method 61–02.01. The peak, trough, final, breakdown, and setback viscosities were measured.

The drying rate, HRYR, and thermal and pasting properties of programs A and B were compared with those from the ordinary drying runs (runs under the ordinary operating conditions of the industry), which maintained a constant grain temperature.

#### 4.1.4 Statistical analysis

The standard deviation was calculated for the HRYR and the drying rates.

Analysis of variance (ANOVA) was used to compare variables. In case of significative difference among variables ($p < 0.05$), Tukey test was applied to determine which were the variables that differ.

### 4.2 Results and discussion

**Table 4** shows the average HRYR and drying rate of the runs dried with program A, program B, and ordinary operating conditions of the industry. Program A and B had significantly higher drying rates than the ordinary runs. In addition, program B had a significantly lower HRYR than the ordinary runs. Although the difference between program A and program B was not significant, it was observed that program B runs tend to have lower HRYR than program A.

Based on these results, program B seems to be the most promising to improve the drying efficiency, increasing the drying rate and reducing the HRYR compared to the ordinary operating conditions. Implementing this program would reduce the drying duration of each run, increasing the reception capacity of the drying plant and, consequently, improving its productivity. At the same time, reducing the HRYR would increase the profitability of the rice obtained.

Drying conditions can affect the quality of rice, especially when high grain temperatures are reached during the drying process [31–33]. For this reason, thermal and pasting properties of the samples from program B and those from the runs under

| | HRYR (pp) | Drying rate (%MC/h) |
|---|---|---|
| Program A | $2.0 \pm 1.0^{A,B}$ | $1.38 \pm 0.2^{A}$ |
| Program B | $0.8 \pm 0.9^{B}$ | $1.28 \pm 0.2^{A}$ |
| OOC | $2.8 \pm 0.4^{A}$ | $1.06 \pm 0.2^{B}$ |

*Different characters in the same column indicate significative difference among samples ($p < 0.05$). Errors correspond to two standard deviations ($\pm 2\sigma$). HRYR Head rice yield reduction and OOC = Ordinary operating conditions.*

**Table 4.**
*HRYR and drying rate of drying programs and ordinary commercial drying runs.*

| | T. Peak (°C) | T. Onset (°C) | T. Conclude (°C) | ΔH (J/g) |
|---|---|---|---|---|
| PROG B-IN | 65.19 | 59.28 | 78.03 | 11.01 |
| PROG B-OUT | 65.52 | 59.74 | 76.19 | 11.56 |
| OOC-IN | 66.06 | 59.95 | 77.89 | 11.65 |
| OOC-OUT | 66.38 | 60.72 | 80.98 | 11.13 |

*Results are the average of two runs in each condition (program B or ordinary operating conditions).*
*PROG, Program; OOC, Ordinary operating conditions; IN, Representative sample of the loading; OUT, Representative sample of the unloading; T, Temperature; ΔH, Enthalpy.*

**Table 5.**
*Thermal properties of rice samples collected during drying.*

ordinary operating conditions were measured and compared. **Table 5** and **Figure 4** show these results.

Comparing the values and profiles of the samples collected during the loading (before drying) with those collected during the unloading (after drying), no significant differences were observed either for samples from Program B or for those from the ordinary operating conditions. The same behavior was observed with samples from program A (data not shown). Therefore, it could be concluded that the drying process had no effect on the cooking properties in any of the conditions tested, confirming that Program B is a suitable drying program to implement in the industry to improve the drying efficiency.

In Ref. (14), the authors found that hardness and stickiness of an aromatic long-grain rice dried using different drying methods were not significantly different to the control sample (sample dried under very mild conditions) when the drying temperature was kept below 60°C. In agreement with this, Dillahunty et al. [32] found that only drying above 55°C with exposure durations higher than 12 hours lowered the peak viscosity of a long-grain (Cypress) and a medium-grain (Bengal) rice. At temperatures below that, there were no significant differences with the control samples. In our experiments, temperature never exceeded 50°C, being below the temperature reported as critical by these researchers. Therefore, our results for a South American variety were in agreement with these findings reported for other long-grain varieties.

## 5. Conclusions

- This chapter reviews the main variables affecting the drying efficiency, understanding the drying efficiency as the combination of drying rate and HRY.

- Drying of a long-grain South American rice variety under controlled drying air conditions was studied using a thin-layer lab-scale dryer. It was found that the HRYR and drying rate were affected by the drying air temperature and RH. Tg also played an important role in the drying process.

- During these experiments, it was concluded that drying at grain temperatures above 47°C without affecting the HRY was possible up to a grain MC of 15%. This allowed an increase in the drying rate (compared to drying at lower

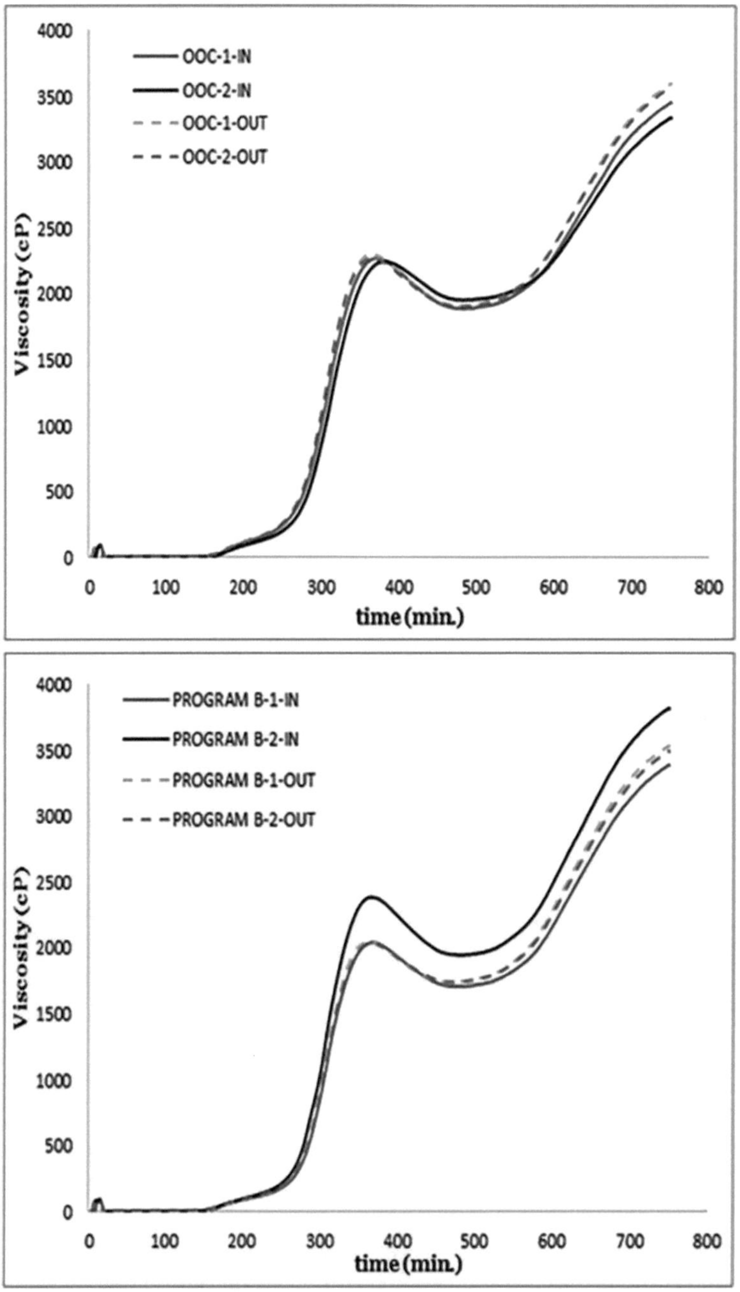


**Figure 4.**
*Viscosity profiles of rice samples collected during drying. OOC, ordinary operating conditions; PROG, program; IN, representative sample of the loading; OUT, representative sample of the unloading.*

temperatures). For MC below 15%, the grain should be dried at milder conditions (lower grain temperatures) to avoid an increase of the HRYR.

- Based on these results, a two-stage drying program was proposed to improve the drying efficiency (drying rate and HRYR). This program was tested in a commercial cross-flow dryer using a South American rice variety with promising results. The HRYR was reduced and the drying rate increased, compared to the runs performed at the ordinary operating conditions of the industry. Additionally, the cooking properties were not affected.

- Implementing this drying program would allow to increase the reception capacity of the rice industries and reduce the percentage of rice kernels with lower added value (broken kernels), improving the productivity of the industrial sector.

## Acknowledgements

We thank LATU (Technological Laboratory of Uruguay) for the financial support.

## Nomenclature

| | |
|---|---|
| EMC | Equilibrium moisture content |
| HRY | head rice yield (pp) |
| HRYR | head rice yield reduction (pp) |
| k | drying rate constant ($h^{-1}$) |
| MC | moisture content, dry basis |
| RH | relative humidity (%) |
| T | temperature (°C) |
| Tg | glass transition temperature (°C) |
| W | weight (g) |
| I | initial |

## Subscript:

| | |
|---|---|
| t | variable at a time t |

## Author details

Laura Garcia-Llobodanin* and Alejandra Billiris
Latitud-LATU Foundation, Montevideo, Uruguay

*Address all correspondence to: lagarcia@latitud.org.uy

## References

[1] Jinsong B. Rice: Chemistry and Technology. Duxford: Elsevier, Inc.; 2019

[2] Chen H, Siebenmorgen TJ, e Marks BP. Relating drying rate constant to head rice yield reduction of long-grain rice. Transactions of the ASAE. 1997;**40**(4): 1133-1139

[3] Buggenhout J et al. The breakage susceptibility of raw and parboiled rice: A review. Journal of Food Engineering. 2013;**117**:304-315

[4] Ondier GO, Siebenmorgen T, e Mauromoustakos A. Drying characteristics and milling quality of rough rice dried in a single pass incorporating glass transition principles. Drying Technology. 2012;**30**: 1821-1830

[5] Cnossen AG et al. An application of glass transition temperature to explain rice kernel fissure occurrence during the drying process. Drying Technology. 2001;**19**(8):1661-1682

[6] Kunze OR, e Choudhury MSU. Moisture adsorption related to the tensile strength of rice. Cereal Chemistry. 1972; **49**(6):684-696

[7] Zeleznak KJ, Hoseney RC. The glass transition in starch. Cereal Chemistry. 1987;**64**(2):121-124

[8] Perdon A, Siebenmorgen TJ, e Mauromoustakos A. Glassy state transition and rice drying: Development of a brown rice state diagram. Cereal Chemistry. 2000;**77**(6):708-713

[9] Cnossen AG, Siebenmorgen TJ, Yang W. The glass transition temperature concept in rice drying and tempering: Effect on drying rate. Transactions of the ASAE. 2002;**45**(3): 759-766

[10] Fan J, Siebenmorgen TJ, Yang W. A study of head rice yield reduction of long- and medium-grain rice varieties in relation to various harvest and drying conditions. Transactions of the ASAE. 2000;**43**(6):1709-1714

[11] Schluterman DA, Siebenmorgen TJ. Relating rough rice moisture content reduction and tempering duration to head rice yield reduction. Transactions of the ASABE. 2007;**50**(1): 137-142

[12] Franco CMR, de Lima AGB, Farias VSO, da Silva WP. Modeling and experimentation of continuous and intermittent drying of rough rice grains. Heat and Mass Transfer. 2020;**56**: 1003-1014

[13] Luthra K, Sadaka S. Investigation of rough rice drying in fixed and fluidized bed dryers utilizing dehumidified air as a drying agent. Drying Technology. 2020; **39**:1059-1073

[14] Inprasit C, e Noomhorm A. Effect of drying air temperature and grain temperature of different types of dryer and operation on rice quality. Drying Technology. 2001;**19**(2): 389-404

[15] Garcia-Llobodanin L, Billiris A. Effect of the drying air conditions on the drying rate and milling quality of a long-grain rice variety. Food Science and Technology. 2023;**43**:1-7

[16] Ondier GO, Siebenmrgen TJ, Mauromoustakos A. Low-temperature, low-relative humidity drying of rough rice. Journal of Food Engineering. 2010; **100**:545-550

[17] Liu Y, Bhandari B, Zhou W. Glass transition and enthalpy relaxation of amorphous food saccharides: A review. Journal of Agricultural and Food Chemistry. 2006;**54**:5701-5707

[18] Liu P et al. Glass transition temperature of starches with different amylose/amylopectin ratios. Journal of Cereal Science. 2010;**51**:388-391

[19] Sun Z et al. Thermomechanical transitions of rice kernels. Cereal Chemistry. 2002;**79**(3):349-353

[20] Garcia-Llobodanin L et al. Effect of rice variety and harvest moisture content on the rice transition temperature of Uruguayan rice varieties. Innotec. 2020; **20**:106-116

[21] Yang W, Jia C. Glass transition mappong inside a rice kernel. Transactions of the ASAE. 2004;**47**(6):2009-2015

[22] Odek Z, Siebenmorgen TJ, Mauromoustakos A. Relative impact of kernel thickness and moisture content on rice fissuring during drying. Applied Engineering in Agriculture. 2018;**34**(1): 239-246

[23] AACC. Method 44-15.02: Moisture air oven methods. In: Approved Methods of Analysis. St. Paul: Cereals & Grains Association; 1999

[24] Ondier GO et al. Equilibrium moisture contents of pureline, hybrid and parboiled rice. Transactions of the ASABE. 2011;**54**(3):1007-1013

[25] Kahveci K, Cihan A, Ece MC. Liquid diffusion model for drying a stack of rough rice. International Journal of Energy Research. 2003;**27**:1141-1143

[26] Pereira JCA et al. Continuous and intermittent drying of rough rice: Effect on process effective time and effective mass diffusivity. Agriculture. 2020; **10**(7):282-294

[27] Hacihafizoğlu O, Cihan A, Kahveci K. Mathematical modeling of drying of thin layer rough rice. Foods and Bioproducts Processing. 2008;**86**: 268-275

[28] Cihan A, Kahveci K, e Hacihafizoğlu O. Modeling of intermittent drying of thin layer rough rice. Journal of Food Engineering. 2007;**79**:293-298

[29] Li X et al. Changes in moisture effective diffusivity and glass transition temperature of paddy during drying. Computers and Electronics in Agriculture. 2016;**128**:112-119

[30] Basutkar NN et al. Functional properties of commingled Rice-cultivar lots. Cereal Chemistry. 2015;**92**(1): 114-119

[31] Graham-Acquaah S, Siebenmorgen TJ. Rice paste viscosities and gel texture resulting from varying drying and tempering regimen. Cereal Chemistry. 2021;**98**(2):285-295

[32] Dillahunty AL, Siebenmorgen TJ, Mauromoustakos A. Effect of temperature, exposure duration, and moisture content on color and viscosity of Rice. Cereal Chemistry. 2001;**78**(5): 559-563

[33] Tirawanichakul S et al. Effect of fluidized bed drying temperature on various quality attributes of Paddy. Drying Technology. 2004;**22**(7): 1731-1754

Section 2

# Drying Equipment

Chapter 4

# Polyphenol Extraction for the Enhancement of Food Lipid Quality, with an Emphasis on the Roles of Extraction Technologies, Moisture and Drying Temperature

*Peter Obasa, Bolanle Adenike Adejumo, James Agajo, Samuel Tunde Olorunsogo and Labake Fadipe*

**Abstract**

Polyphenols increase food lipid quality, the taste, stability, and health advantages of lipids in different dietary applications. Polyphenol content depends on the extraction process, moisture, and drying temperature. Polyphenol due to antioxidant and antibacterial capabilities, natural compounds, are used to improve dietary lipid quality. However, polyphenol extraction has been a very challenging task that has caused drawback in the fortification of food lipid. Extraction process of polyphenol and solvent, solid-phase, and supercritical fluid extraction techniques has been analysed. Moisture and drying temperature affect extraction efficiency quality. Optimised polyphenol extraction in the connections between polyphenols, extraction technique, moisture, and drying temperature needs to be more examined. Polyphenols role in dietary lipid quality is discussed, and food source polyphenol content needs to be well researched. Drying temperature impacts extraction efficiency as it was carried out in previous research, and moisture content affects polyphenol solubility. Polyphenol extraction improves lipid quality in olive oil enrichment, meat, poultry, dairy, nutritional supplements, and bread and confectionery goods. Stability, extraction efficiency, selectivity, standardisation, sustainability, and industrial adoption are still issues. Stability, preservation, green extraction, and industrial scalability are research priorities. Complicated interactions need to be understood for future research directions in polyphenol extraction for lipid quality enhancement.

**Keywords:** extraction, lipid, quality, storability, ultrasonic

## 1. Introduction

Food lipid quality plays an essential role in determining the sensory attributes, nutritional content, and shelf life of various food products. Lipids are a concentrated source of energy and vital fatty acids and enhance the flavour, texture, and fragrance of food. However, lipids can deteriorate due to oxidation, which can cause rancidity, bad flavours, and nutritional loss. Therefore, it is essential to maintain and improve the quality of food lipids in order to satisfy consumers and advance general health.

Recently, there has been a rise in interest in using natural substances to enhance the quality of dietary lipids. Among these substances, polyphenols have emerged as important bioactive elements known for strong antioxidant and antibacterial capabilities. Because of capacity to scavenge free radicals and suppress oxidative processes, polyphenols which are widely prevalent in fruits, vegetables, nuts, and other plant-based sources have been recognised for positive impacts on human health [1, 2].

To carefully examine the extraction methods used to extract polyphenols, such as solvent extraction, solid-phase extraction, and supercritical fluid extraction, in order to provide a thorough understanding [3, 4]. Examine at how moisture and drying temperature affect polyphenol extraction effectiveness and how that affects the quality of food lipids [5, 6]. Will also look at how these variables interact and how it affects lipid quality improvement and the efficiency of polyphenol extraction [7, 8].

For the extraction process to be optimised and the full potential of polyphenols to improve the quality of food lipids to be realised, it is essential to comprehend the complex interactions between polyphenols, extraction technology, moisture, and drying temperature. This chapter intends to shed vital light on these linkages for researchers, food scientists, and industry experts interested in using polyphenols as natural additives to enhance lipid quality.

### 1.1 Definition of polyphenols and importance in food lipid quality improvement

Polyphenols are a broad class of bioactive compounds found abundantly throughout different plant-based sources of nutrition, including fruits, vegetables, whole grains, legumes, and herbs. These substances have drawn a lot of interest because of possible health advantages and contribution to improving food lipid quality. The definition of polyphenols, significance in enhancing the quality of dietary lipids, and the interaction between moisture, drying temperature, and extraction method in polyphenol extraction from food sources will all be covered in this book chapter.

## 2. Overview of the role of moisture, drying temperature, and extraction technology in polyphenol extraction from food sources

The presence of numerous phenolic groups distinguishes polyphenols, a general family of secondary metabolites found in plants. Flavonoids (flavones, flavonols, flavanones, etc.), phenolic acids (hydroxybenzoic acids, hydroxycinnamic acids), stilbenes, lignans, and tannins are some of the subclasses into which these chemicals are divided (**Figure 1**). According to Perino-Issartier et al. [10], each subclass has distinct biological and chemical characteristics. The Importance of Polyphenols in the Improvement of Food Lipid Quality Polyphenols have been thoroughly investigated for ability to improve food lipid quality. They have exceptional antioxidant qualities that can prevent lipid oxidation, which is the main factor in the rancidity and degeneration of lipid-based foods. Lipid oxidation causes sensory alterations, such as off flavours and colour deterioration,

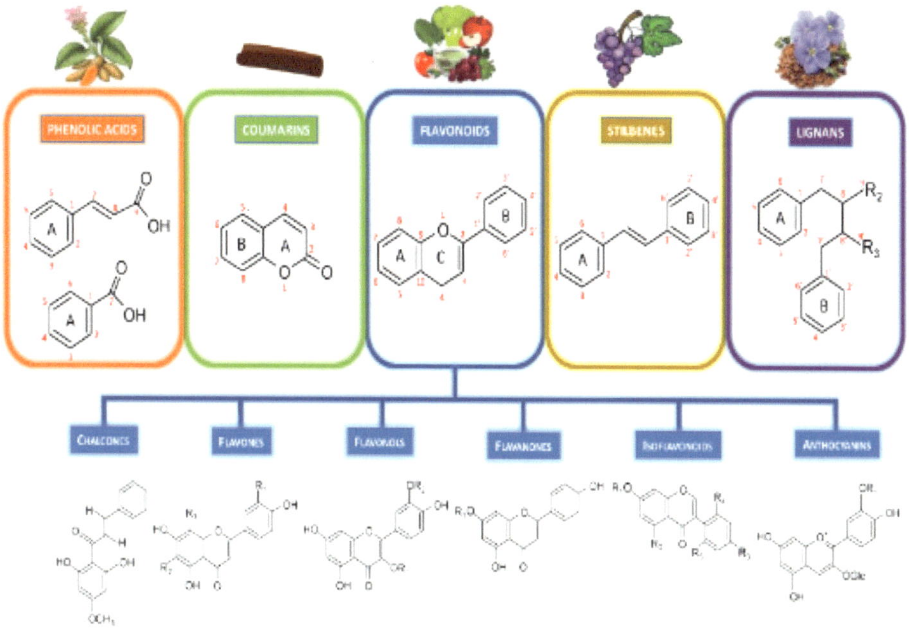


**Figure 1.**
*Plant polyphenol classification (source: Golmakani et al. [9]).*

as well as the loss of vitamins and important fatty acids, which lowers the food's nutritional value [11]. In order to successfully prevent oxidative damage and increase the shelf life of foods containing lipids, polyphenols work as free radical scavengers, chelators of pro-oxidant metal ions, and regulators of lipid peroxidation [12].

### 2.1 Moisture and polyphenol extraction

The amount of moisture in raw materials has a big influence on how well polyphenols are extracted. Water can make polyphenols more soluble, facilitating release from the food matrix during extraction. However, too much moisture can also cause enzymatic and non-enzymatic processes that cause polyphenols to degrade and disappear. In order to maximise the yield of polyphenol extraction and maintain the quality of the end product, moisture content must be well controlled [13].

Drying Temperature and Polyphenol Extraction: Another important factor that affects polyphenol extraction is drying temperature. By making polyphenols more solubilised and diffusible, higher temperatures can quicken the extraction process. However, too much heat can also result in bioactivity loss and thermal deterioration. In order to maximise polyphenol extraction efficiency while maintaining veracity, it is crucial to determine the ideal drying temperature [6, 7].

### 2.2 Polyphenol extraction and extraction technology

To effectively recover polyphenols from dietary sources, a variety of extraction processes are used. Common techniques include solvent extraction, solid-phase extraction, and supercritical fluid extraction (**Figure 2**). Solid-phase extraction

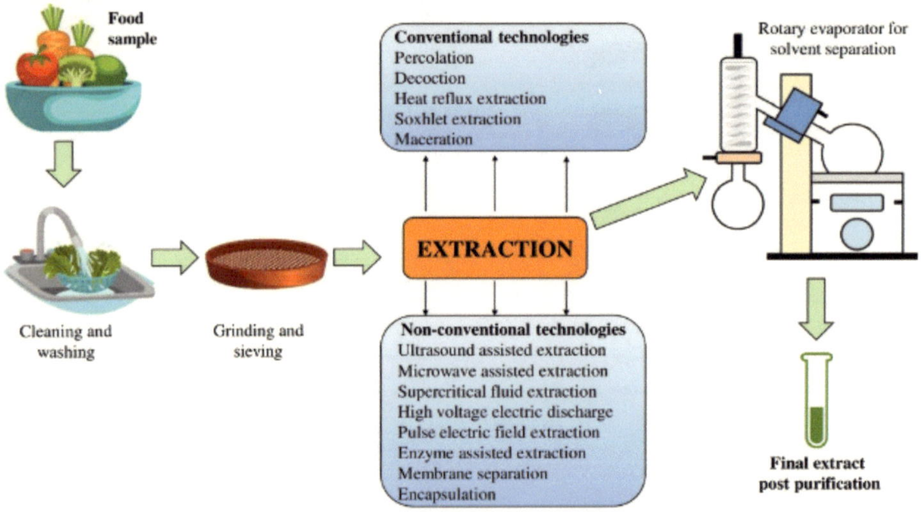


**Figure 2.**
*Polyphenol extraction techniques and methods (source Sridhar et al. [14]).*

makes use of solid adsorbents to selectively capture target chemicals, whereas solvent extraction uses organic solvents to extract polyphenols.

Polyphenols are extracted using supercritical fluids, such as carbon dioxide, under carefully regulated circumstances. The extraction yield, selectivity, and cost-effectiveness are all impacted by the benefits and limits of each approach [5, 15, 16]. The antioxidant abilities, how they affect sensory qualities, and how they can make lipid-based foods last longer on store shelves. How moisture content and drying temperature affect the effectiveness of polyphenol extraction and the preservation of bioactivity. For the [17] extraction process to be optimised and the full potential of polyphenols to enhance food lipid quality to be realised, it is essential to comprehend how moisture, drying temperature, and extraction technique interact. This chapter tries to clarify these interactions in order to offer helpful information to researchers, food scientists, and business experts who are considering using polyphenols as all-natural substances to improve lipid quality.

## 2.3 Factors affecting polyphenol content in food sources

Several factors, including genetic variation, agricultural practises, environmental conditions, and post-harvest processing, influence the polyphenol content of food sources. For optimising polyphenol extraction and ensuring the quality of polyphenol-rich products, it is essential to comprehend these factors.

### *2.3.1 Genetic variation*

The polyphenol content of various plant varieties or cultivars varies. Genetic factors play an important role in determining the types and amounts of polyphenols present in different plant species [18].

#### *2.3.2 Agricultural practises*

Agricultural practises, such as fertilisation, irrigation, and pesticide application, can affect the polyphenol content of food sources. Montesano et al. [19] report that organic farming methods and sustainable agricultural practises increase polyphenol levels compared to conventional methods. The environment, including climate, exposure to sunlight, and soil composition, can have an effect on polyphenol synthesis in plants. As a defence mechanism against oxidative stress, increased ultraviolet (UV) radiation exposure can stimulate the synthesis of certain polyphenols [20].

## 3. Influence of moisture content on the extraction of polyphenols from food sources

Moisture content serves a crucial function in the extraction of polyphenols from food sources. It influences the extraction process's solubility, stability, and release of polyphenols. Understanding the effect of moisture content is crucial for optimising extraction efficacy and maintaining polyphenol integrity.

### 3.1 Solubility and diffusion

According to Kahraman et al. [21], moisture can increase the solubility and diffusion of polyphenols from the food matrix into the extraction solvent, resulting in a higher extraction efficiency.

### 3.2 Enzymatic reactions

Excessive moisture content can cause enzymatic reactions, such as polyphenol oxidase activity, which result in the degradation and loss of polyphenols during extraction [21].

## 4. Mechanisms of polyphenol degradation and oxidation under various conditions of humidity

Polyphenols are susceptible to degradation and oxidation under certain conditions of moisture. The presence of water can initiate chemical reactions that compromise polyphenols' stability and bioactivity. Moisture can catalyse hydrolysis reactions, resulting in the dissolution of glycosidic bonds and the release of aglycones. This procedure can alter polyphenols' composition and bioactivity.

Moisture can stimulate oxidative reactions, leading to the oxidation of polyphenols and the production of reactive oxygen species. These reactions can reduce polyphenols' antioxidant capacity [12, 21].

### 4.1 Moisture control strategies for polyphenol extraction

Drying and Pre-treatment Methods: Controlling the moisture content is crucial for optimising the extraction of polyphenols. Various techniques, such as dehydrating and pretreatment procedures, can be used to control moisture levels and improve extraction efficacy.

Drying is frequently employed to minimise the moisture content of dietary sources prior to extraction. Freeze-drying, air drying, and microwave drying are all effective techniques for removing moisture while maintaining polyphenol integrity [22].

Pre-treatment Methods: Pre-treatment methods, such as blanching, steaming, and marinating, can be used to modify the moisture content and structure of food sources, thereby facilitating the release of polyphenols during extraction [13].

## 5. Examples of food applications utilising polyphenol extraction for lipid quality enhancement

Various food applications have effectively utilised polyphenol extraction to improve lipid quality. These applications demonstrate the ability of polyphenols to enhance the sensory qualities, stability, and health-promoting properties of lipids. The following instances illustrate the efficacy of polyphenol extraction in various food contexts:

Olive Oil Enrichment Olive oil has been enriched with polyphenol-rich compounds derived from olive pomace or leaves. The incorporation of these extracts increases the lipid content of the oil by enhancing its oxidative stability and antioxidant capacity [19, 23].

Incorporating polyphenol extracts into meat and poultry formulations can improve lipid stability and prevent lipid oxidation [24, 25]. These extracts function as natural antioxidants, preventing the formation of rancid flavours and preserving the overall quality of the products.

To enhance the lipid quality of dairy products, polyphenol extracts from sources such as grape seeds and berry pomace have been added [26, 27]. These extracts can improve the oxidative stability and inhibit the development of off-flavours in dairy products such as yoghurt, cheese, and butter.

Polyphenol-rich extracts have been incorporated into the formulation of nutritional supplements, particularly those that target lipid metabolism and cardiovascular health. These extracts, derived from green tea, cocoa, or citrus fruits, contribute to the positive effects of supplements on lipid profiles and oxidative stress markers [28, 29].

Bakery and Confectionery Products: Polyphenol compounds have been added to bakery and confectionery products to improve lipid quality and extend storage life [30, 31]. These extracts can inhibit lipid oxidation and enhance overall oxidative stability, thereby preserving the sensory attributes and freshness of the products.

The impact of dietary polyphenols on gut microbiota compositions and how gut microbiota affect polyphenol metabolism are the main possible mechanisms of the gut microbiome. In combination, dietary polyphenols could be a helpful nutritional strategy for modifying lipid metabolism or treating obesity [32]. According to studies by [33], dietary TP appears to have had a major role in activating lipolysis and enhancing lipid metabolism in grouper given high-fat diets. For enhancing lipid metabolism in juvenile grouper high-lipid diets, 0.045–0.067% was advised based on the broken-line regressions analysis of the serous LDL and hepatic VLDL contents [34]. The effects of TP concentration (0, 100, and 400 mg/L), solution pH (3 and 7), and the addition of ethylenediaminetetraacetic acid (0 or 20 mol/L EDTA) were investigated on the antioxidant/prooxidant activities of TP in walnut oil-in-water

(O/W) emulsions under accelerated storage conditions (50°C for 0, 48, and 96 h). At pH 3, TP concentrations of 100 mg/L and 400 mg/L both showed better antioxidant activity for lipids and proteins, respectively.

## 6. Challenges and future perspectives polyphenol extraction

Despite the significant advancements in polyphenol extraction for augmenting food lipid quality, a number of obstacles and opportunities exist for future research and application. This section discusses a few of these obstacles and offers prospective future perspectives.

### 6.1 Stability and preservation

The stability of polyphenols during processing, storage, and in the presence of other food components is one of the greatest obstacles to utilising polyphenols to enhance lipid quality. Polyphenols are susceptible to degradation and oxidation, which result in diminished efficacy and bioactivity. Strategies to improve polyphenol stability, such as encapsulation techniques and the development of protective delivery systems, require investigation [35, 36].

Although numerous extraction techniques have been devised, achieving high extraction yield and efficiency remains a challenge. Optimal polyphenol extraction necessitates the optimization of extraction parameters, the selection of suitable solvents, and the investigation of innovative extraction techniques [37].

The selective extraction of particular polyphenols from complex food matrices is a challenge that must be addressed. Different polyphenols exhibit distinct bioactivities, and the extraction selectivity of these polyphenols can affect the final product's desired functional properties. It is of interest to develop extraction methods that target specific polyphenols while preserving integrity [27, 37].

### 6.2 Standardisation and quality control

Standardisation of polyphenol extraction protocols and the establishment of quality control measures are required to ensure consistent and reliable extraction results. The development of analytical methods and the establishment of quality standards will contribute to the reproducibility and comparability of research findings [38].

### 6.3 Sustainability and green technologies

As the demand for sustainable and environmentally benign practices increases, it is necessary to develop green extraction technologies that minimise energy consumption, solvent usage, and refuse production. The investigation of alternative solvents, such as supercritical fluids and natural deep eutectic solvents, can provide environmentally preferable alternatives for polyphenol extraction [39].

### 6.4 Industrial implementation and scalability

Although polyphenol extraction techniques have demonstrated promise in laboratory studies, implementation on an industrial scale poses challenge [40]. Scaling up the extraction process while maintaining its efficacy, cost-effectiveness, and product quality requires additional research and development.

Future research need to concentrate on addressing these obstacles and investigating new opportunities in polyphenol extraction for the enhancement of lipid quality. The development of advanced extraction technologies, the investigation of synergistic effects of polyphenols with other bioactive compounds, and the application of computational modelling and artificial intelligence to optimise extraction processes are potential future research directions [30].

Despite the progress made in the extraction of polyphenols to improve the lipid quality of foods, several challenges and future perspectives remain. Key areas requiring attention include stability and preservation, extraction efficiency, selectivity and specificity, standardisation and quality control, sustainability and green technologies, industrial implementation, and scalability. Addressing these obstacles and investigating new research avenues will pave the way for advances in polyphenol extraction techniques and contribute to the development of lipid-based food products with enhanced health benefits.

## 6.5 Challenges

While the extraction of polyphenols for improving food lipid quality holds great promise, there are several challenges that need to be addressed. One major challenge is the variability of polyphenol content in different food sources. The composition and concentration of polyphenols can vary significantly, making it challenging to establish standardised extraction protocols. Additionally, the stability of polyphenols during extraction and processing can be compromised due to sensitivity to environmental factors, such as light, heat, and pH.

Another challenge is the selection of appropriate extraction technologies. Various extraction methods, including conventional techniques (solvent extraction) and emerging green technologies (supercritical fluid extraction, ultrasound-assisted extraction), are available. However, the choice of the most suitable extraction method depends on factors such as the target polyphenol, the food matrix, the desired yield, and the cost-effectiveness of the process. Optimization of extraction conditions is crucial to achieve high extraction efficiency and preserve the integrity of polyphenols.

Furthermore, the scale-up of polyphenol extraction processes from lab-scale to industrial scale presents challenges. Industrial production requires higher through put and cost-effective operations, while maintaining the quality and bioactivity of extracted polyphenols. Ensuring scalability and reproducibility while meeting regulatory requirements can be demanding.

## 6.6 Future perspectives

Despite the challenges, there are several exciting future research directions that can advance the field of polyphenol extraction and its application in enhancing food lipid quality.

One area of interest is the development of novel extraction techniques that offer improved efficiency, selectivity, and sustainability. Techniques such as microwave-assisted extraction, enzyme-assisted extraction, and pulsed electric field extraction show promise in enhancing polyphenol extraction yields and reducing processing time and solvent consumption.

The utilisation of by-products and waste materials as potential sources of polyphenols is gaining attention. Researchers are exploring the valorisation of agricultural residues, food processing by-products, and plant waste materials for the extraction of polyphenols. This approach not only reduces waste but also offers economic and environmental benefits.

The application of advanced analytical techniques for the characterisation of polyphenols is another avenue for future research. High-performance liquid chromatography (HPLC), gas chromatography–mass spectrometry (GC–MS), and nuclear magnetic resonance (NMR) spectroscopy enable the identification and quantification of specific polyphenols and metabolites, providing insights into bioavailability and potential health benefits.

Furthermore, the development of encapsulation technologies for polyphenols can improve stability and controlled release in food products. Encapsulation techniques, such as spray drying, coacervation, and Nano emulsion, protect polyphenols from degradation and enhance functionality in food matrices.

In the challenges related to polyphenol extraction, such as variability in polyphenol content, selection of suitable extraction technologies, and scalability, is essential for the successful implementation of polyphenols in improving food lipid quality. Future research efforts should focus on the development of novel extraction techniques, utilisation of by-products, advanced analytical characterisation, and encapsulation technologies. By overcoming these challenges and exploring new opportunities, the field of polyphenol extraction can advance, leading to innovative and sustainable approaches for enhancing food lipid quality.

## 7. Future research directions in polyphenol extraction for lipid quality enhancement

The goal of ongoing research into polyphenol extraction is to further enhance the process's efficacy, sustainability, and scalability. The following are possible directions for future research in this field:

Enzyme-assisted extraction, intermittent electric fields, and pressurised liquid extraction are a few of the novel extraction techniques being investigated by scientists. These techniques offer benefits such as reduced extraction time, increased yield, and improved bioavailability of polyphenols [41, 42].

Further optimisation of extraction parameters, such as solvent selection, extraction time, temperature, and pressure, can assist in maximising polyphenol yield and quality. Design of experiments (DoE) and response surface methodology (RSM) are important optimisation tools for these extraction procedures [28, 38].

### 7.1 Advances in encapsulation techniques

Encapsulation of polyphenols can enhance stability and bioavailability, allowing them to be incorporated into a wider array of lipid-based products. For the development of effective delivery systems, nanoparticles, microencapsulation, and emulsion based systems are of interest [27, 39].

### 7.2 Exploration of underutilised sources

Research into underutilised plant sources abundant in polyphenols can increase the availability of renewable resources for extraction. By utilising these sources, researchers can discover novel polyphenols with distinct properties and optimise extraction techniques [30, 40].

### 7.3 Considerations regarding the sustainability and scalability of polyphenol extraction

As the demand for polyphenols rises, it is crucial that extraction processes consider sustainability and scalability, particularly for industrial applications. Important factors include:

The development of environmentally favourable extraction techniques that minimise solvent usage, energy consumption, and refuse production is crucial. Bashari et al. [41] and Chiorcea-Paquim et al. [42] report that subcritical water extraction, solid phase extraction, and green solvents contribute to the sustainability of polyphenol extraction.

Utilising by-products generated during the extraction process, such as fruit husks, pomace, and seed residues, reduces waste and improves the economic viability of extraction. The overall sustainability of the process can be improved by extracting polyphenols from these by-products [43, 44].

Life Cycle Assessment: Conducting life cycle assessments (LCAs) of polyphenol extraction processes allows for the evaluation of environmental impacts throughout the entire lifecycle, from the production of primary materials to extraction and refuse management. LCAs provide insightful information for optimising processes and making informed decisions [45, 46].

## 8. Conclusion

This chapter examines the definition of polyphenols and significance in enhancing the lipid quality of food. It has investigated the role of moisture, dehydrating temperature, and extraction technology in the extraction of polyphenols from food sources. In addition, the factors influencing polyphenol content, the effect of moisture on extraction, the mechanisms of polyphenol degradation, strategies for moisture control, and the intricate interactions between polyphenols, extraction technology, moisture, and drying temperature. In addition, examples of effective food applications in which polyphenol extraction was used to improve lipid quality.

The primary findings of this chapter emphasise the significant impact of polyphenol extraction techniques on the need for improvement as limited research finding are available. Polyphenols have been shown to enhance the sensory qualities, stability, and health-promoting properties of lipids in a variety of food applications. By incorporating polyphenol extracts into olive oil, meat and poultry products, dairy products, dietary supplements, and bakery and confectionary products, lipid stability and oxidative stability can be improved, resulting in enhanced overall quality and a longer shelf life.

In addition, this chapter has illuminated the significance of moisture, dehydrating temperature, and extraction technology in polyphenol extraction processes. Controlling the moisture content is essential for preventing polyphenol degradation and oxidation, as it plays a vital role in extraction efficiency. Moisture can be effectively controlled and polyphenol extraction optimised using techniques such as dehydration and pre-treatment.

Future research directions in polyphenol extraction for the enhancement of lipid quality include the development of novel extraction techniques, optimisation strategies, advances in encapsulation techniques, and the exploration of underutilised sources. These developments seek to improve the efficacy, sustainability, and scalability of polyphenol extraction processes, resulting in higher extraction yields and higher-quality polyphenols for a variety of applications.

The extraction of polyphenols has significant implications for the food industry. The incorporation of polyphenol-rich extracts into food products is a natural and sustainable method for improving lipid quality, oxidative stability, and lipid oxidation inhibition. This not only contributes to the sensory qualities and expiration life of food products but also provides consumers with potential health benefits.

By utilising the power of polyphenols, the food industry can enhance the quality and stability of lipid-based products, thereby providing consumers with healthier and more appetising options.

## Author details

Peter Obasa[1*], Bolanle Adenike Adejumo[1], James Agajo[2], Samuel Tunde Olorunsogo[1] and Labake Fadipe[3]

1 Agricultural and Bioresources Engineering, Federal University of Technology Minna, Minna, Nigeria

2 Computer Engineering, Federal University of Technology Minna, Minna, Nigeria

3 Chemistry Department, Federal University of Technology Minna, Minna, Nigeria

*Address all correspondence to: peter.obasa@futminna.edu.ng

## References

[1] Garcia-Lomillo J, Gonzalez-SanJosé ML, Del Pino-García R, Ortega-Heras M, Muñiz-Rodríguez P. Polyphenol consumption for reducing oxidative stress in the human body: A systematic review. Plant Foods for Human Nutrition. 2017;**72**(4):400-416. DOI: 10.1007/s11130-017-0634-8

[2] Gulçin I. Antioxidant activity of food constituents: An overview. Archives of Toxicology. 2012;**86**(3):345-391. DOI: 10.1007/s00204-011-0774-2

[3] Boulekbache-Makhlouf L, Medouni L, Medouni-Adrar S, Arkoub L. Effect of extraction solvent on phenolic compounds content and antioxidant activities from the leaves of Algerian Tetraclinis articulata (Vahl) masters. Industrial Crops and Products. 2013;**41**:85-91. DOI: 10.1016/j.indcrop.2012.03.009

[4] Khadra I, Tavazzi I, Koubaa M, Barba FJ, Mhemdi H, Vorobiev E, et al. An insight into the role of non-conventional extraction techniques assisted by sustainable energies for enhanced recovery of phenolic compounds from Agri-food by-products. Food Research International. 2019;**123**:467-483. DOI: 10.1016/j.foodres.2019.04.027

[5] Rezzoug M, Spigno G, Meca G, Ramesh V, Myc A, Ferchichi A. Impact of solvent extraction and supercritical fluid extraction on the recovery of bioactive compounds from plant matrices. Food. 2020;**9**(1):70. DOI: 10.3390/foods9010070

[6] Chemat F, Rombaut N, Meullemiestre A, Turk M, Perino S, Fabiano-Tixier AS. Review of green food processing techniques. Preservation, transformation, and extraction. Innovative Food Science & Emerging Technologies. 2019;**41**:357-377. DOI: 10.1016/j.ifset.2017.05.019

[7] Radocaj O, Dimić E, Tsao R, Tsao R. Antioxidant and antimicrobial polyphenols from soybean waste obtained by ultrasound-assisted extraction. Food Chemistry. 2017;**218**:14-21. DOI: 10.1016/j.foodchem.2016.08.069

[8] Obasa Peter A, Adejumo Bolanle A, James A, Samuel O. Development of adaptive biosensor system approach for food products in storage (a review). Iconic Research and Engineering Journals. 2023;**6**(10):721-727. Available from: https://www.irejournals.com/paper-details/1704220

[9] Golmakani MT, Rezaei K, Mazidi S, Mortazavi SA. Effect of microwave-assisted extraction on the extraction efficiency of phenolic compounds from olive leaves (Olea europaea L.). Food Chemistry. 2018;**252**:142-148. DOI: 10.1016/j.foodchem.2018.01.062

[10] Perino-Issartier S, Combes D, Dornier M. Role of mass transfer in polyphenol extraction from grape by-products during thermal treatment. Journal of Food Engineering. 2019;**243**:101-110. DOI: 10.1016/j.jfoodeng.2018.09.018

[11] Hano C, Tungmunnithum D. Plant polyphenols, more than just simple natural antioxidants: Oxidative stress, aging and age-related diseases. Medicine. 2020;**7**(5):26. DOI: 10.3390/medicines7050026

[12] Adejumo BA, Ajibo RU, Obasa PA. Optimization of soaking condition and drying temperature for the

production of African yam beans (Sphenostylis sternocarpa) flour. Food Security and Safety: African Perspectives. 2021;**3**(1):63-95. DOI: 10.1007/978-3-030-50672-8_4

[13] Saura-Calixto F. Dietary fiber as a carrier of dietary antioxidants: An essential physiological function. Journal of Agricultural and Food Chemistry. 2011;**59**(1):43-49. DOI: 10.1021/jf103434b

[14] Sridhar A, Ponnuchamy M, Kumar PS. Techniques and modeling of polyphenol extraction from food: A review. Environmental Chemistry Letters. 2021;**19**:3409-3443. DOI: 10.1007/s10311-021-01217-8

[15] Wojdyo A, Oszmiański J, Czemerys R. Antioxidant activity and phenolic compounds in 32 selected herbs. Food Chemistry. 2007;**105**(3):940-949. DOI: 10.1016/j.foodchem.2007.04.038

[16] Neekhra S, Pandith JA, Mir NA, Manzoor A, Ahmad S, Ahmad R, et al. Innovative approaches for microencapsulating bioactive compounds and probiotics: An updated review. Journal of Food Processing and Preservation. 2022;**46**(11):e16935. DOI: 10.1111/jfpp.16935

[17] Siyanbola AA, Obasa PA, Shaba IM. Extraction and optimization of Azardirachta indica seed oil for biodiesel production. 2020;**44**(1):13-15. Available from: https://www.researchgate.net/publication/340088938_Extraction_and_Optimization_of_Azardirachta_Indica_Seed_Oil_for_Biodiesel

[18] Palmieri L, Masuero D, Martinatti P, Baratto G, Martens S, Vrhovsek U. Genotype-by-environment effect on bioactive compounds in strawberry (Fragaria x ananassa Duch.). Journal of the Science of Food and Agriculture. 2017;**97**(12):4180-4189. DOI: 10.1002/jsfa.8290

[19] Montesano V, Negro D, Sonnante G, Laghetti G, Urbano M. Polyphenolic compound variation in globe artichoke cultivars as affected by fertilization and biostimulants application. Plants. 2022;**11**(15):2067. DOI: 10.3390/plants11152067

[20] TungmunnithumD,ThongboonyouA, Pholboon A, Yangsabai A. Flavonoids and other phenolic compounds from medicinal plants for pharmaceutical and medical aspects: An overview. Medicine. 2018;**5**(3):93. DOI: 10.3390/medicines5030093

[21] Kahraman O, Malvandi A, Vargas L, Feng H. Drying characteristics and quality attributes of apple slices dried by a non-thermal ultrasonic contact drying method. Ultrasonics Sonochemistry. 2021;**73**:105. DOI: 10.1016/j.ultsonch.2021.105510

[22] Romero C, Medina E, Mateo MA, Brenes M, de Castro A. In vitro activity of olive oil polyphenols against helicobacter pylori. Journal of Agricultural and Food Chemistry. 2018;**66**(11):2578-2585. DOI: 10.1021/acs.jafc.7b05868

[23] Pan X, Ma G, Fang Y. Tea polyphenols inhibit the formation of atherosclerosis in apolipoprotein E-deficient mice: A bibliometric analysis and systematic review. Frontiers in Pharmacology. 2019;**10**:494. DOI: 10.3389/fphar.2019.00494

[24] Portu J, Gutiérrez-Viguera AR, Gonzalez-Arenzana L, Santamaria P. Characterization of the color parameters and monomeric phenolic composition of 'Tempranillo' and 'Graciano' wines made by carbonic maceration. Food Chemistry. 2023;**406**:155-172. DOI: 10.1016/j.foodchem.2022.134327

[25] Hernandez-Hernandez C, Morales-Sillero A, Fernandez-Prior MÁ, Fernandez-Bolanos J, de la Paz Aguilera-Herrera M, Rodríguez-Gutiérrez G. Extra virgin olive oil jam enriched with cocoa bean husk extract rich in theobromine and phenols. Lebensmittel-Wissenschaft & Technologie. 2019;**111**:278-283. DOI: 10.1016/j.lwt. 2019.05.027

[26] Perez-Jimenez J, Ruano J, Perez-Martínez P, Lopez-Segura F, Lopez-Miranda J. The influence of olive oil on human health: Not a question of fat alone. Molecular Nutrition & Food Research. 2020;**64**(2):2000074. DOI: 10.1002/mnfr.202000074

[27] Jawaid S, Talpur FN, Sherazi TH. Microencapsulation of bioactive compounds: Encapsulation techniques and food applications. Food Hydrocolloids. 2019;**89**:839-850. DOI: 10.1016/j.foodhyd.2018.12.046

[28] Barba FJ, Zhu Z, Koubaa M, Sant'Ana AS, Orlien V. Green alternative methods for the extraction of antioxidant bioactive compounds from winery wastes and by-products: A review. Trends in Food Science & Technology. 2016;**49**:96-109. DOI: 10.1016/j.tifs.2016.01.002

[29] Liang N, Kitts DD, Yu C. Optimization of polyphenol extraction from blueberry leaves (Vaccinium corymbosum L.) using response surface methodology. Food Chemistry. 2020;**318**:126468. DOI: 10.1016/j. foodchem.2020.126468

[30] Ghribi AM, Goula AM, Vorobiev E. Polyphenols recovery from food by-products using green technologies: Advances and perspectives. Critical Reviews in Food Science and Nutrition. 2021;**61**(3):384-406. DOI: 10.1080/10408398.2020.1784 397

[31] Ozcan T, Akpinar-Bayizit A, Yilmaz-Ersan L, Delikanli B, Pekmez H. Extraction of phenolic compounds: A review. Current Analytical Chemistry. 2020;**16**(3):230-245. DOI: 10.2174/ 1573411015666190724151202

[32] Ma J, Zheng Y, Tang W, Yan W, Nie H, Fang J, et al. Dietary polyphenols in lipid metabolism: A role of gut microbiome. Animal Nutrition. 2020;**6**(4):404-409. DOI: 10.1016/j. aninu.2020.08.002

[33] Pan S, Yan X, Li T, Suo X, Liu H, Tan B, et al. Tea polyphenols improve lipid metabolism in hybrid grouper (Epinephelus fuscoguttatus E. lanceolatu) fed high-lipid diets. Aquaculture Reports. 2022;**25**:101217

[34] Tian L, Zhang S, Yi J, Zhu Z, Cui L, Decker EA, et al. Factors impacting the antioxidant/prooxidant activity of tea polyphenols on lipids and proteins in oil-in-water emulsions. Lebensmittel-Wissenschaft & Technologie. 2022;**156**:113024

[35] Sanchez-Rodriguez L, Martin-Escolano R, Fernandez-Arteaga A, Fernandez-Gutierrez A. Deep eutectic solvents: Green alternative for polyphenols extraction from food matrices. Trends in Food Science & Technology. 2021;**112**:678-692. DOI: 10.1016/j.tifs.2021.07.011

[36] Tang M, Zhang M, Liu F, Zhang J, Yang H. Extraction, purification, and industrialization of bioactive substances from plants: Recent advances and challenges. Journal of Separation Science. 2019;**42**(18):2964-2986. DOI: 10.1002/ jssc.2019 00097

[37] Esclapez MD, Garcia-Perez JV, Mulet A. Potential applications of ultrasound in the extraction process of bioactive compounds from plant

materials. Food Research International. 2017;**100**(Part 1):40-57. DOI: 10.1016/j.foodres.2017.07.034

[38] Tourlousse DM, Narita K, Miura T, Ohashi A, Matsuda M, Ohyama Y, et al. Characterization and demonstration of mock communities as control reagents for accurate human microbiome community measurements. Microbiology Spectrum. 2022;**10**(2):e01915-e01921. DOI: 10.1128/spectrum.01915-21

[39] Ozcan O, Temelli F, Goguş F. Supercritical carbon dioxide extraction of polyphenols from pomegranate (Punica granatum L.) peel: Optimization and comparative studies with conventional methods. Separation Science and Technology. 2020;**55**(3):595-607. DOI: 10.1080/01496395.2019.1665526

[40] Sanchez-Rodríguez L, de la Caba K, Peponi L. Nanocomposites based on polyphenols: From extraction to application. Molecules. 2021;**26**(6):1637. DOI: 10.3390/molecules26061637

[41] Bashari M, Moghaddam HN, Ramakrishnan Y, Karimi K. A review on green extraction techniques for polyphenols from plant matrices: State of the art. ACS Sustainable Chemistry & Engineering. 2021;**9**(8):2922-2946. DOI: 10.1021/acssusche meng.0c07785

[42] Chiorcea-Paquim AM, Enache TA, De Souza Gil E, Oliveira-Brett AM. Natural phenolic antioxidants electrochemistry: Towards a new food science methodology. Comprehensive Reviews in Food Science and Food Safety. 2020;**19**(4):1680-1726. DOI: 10.1111/1541-4337.12566

[43] Barba FJ, Boussetta N, Vorobiev E. Emerging technologies for the recovery of isothiocyanates, protein and phenolic compounds from rapeseed and rapeseed press-cake: Effect of high voltage electrical discharges. Innovative Food Science & Emerging Technologies. 2015;**31**:67-72. DOI: 10.1016/j.ifset.2015.06.008

[44] da Silva LB, de Carvalho MG, Amorim AC. Valorization of fruit by-products as a source of bioactive compounds: Extraction and applications. Food Research International. 2021;**144**:110345. DOI: 10.1016/j.foodres.2021.110345

[45] Bekhet HA, Salyer SA. Sustainability of polyphenol extraction from pomegranate peel using life cycle assessment and techno-economic analysis. Science of the Total Environment. 2021;**750**:141542. DOI: 10.1016/j.scitotenv.2020.141542

[46] Ceriani R, Grisenti P, Sala S, Buttol P. Life cycle assessment of polyphenols extraction from spent coffee grounds. Science of the Total Environment. 2020;**705**:135-149. DOI: 10.1016/j.scitotenv.2019.135879

Chapter 5

# Perspective Chapter: Pharmaceutical Drying

*Sachin Kothawade, Vishal Pande, Vaibhav Wagh, Kalyani Autade, Sandesh Bole, Rajashri Sumbe and Shubhangi Albhar*

## Abstract

This chapter presents an overview of the perspective chapter on pharmaceutical drying within the context of drug manufacturing. It explores the significance of pharmaceutical drying in ensuring the stability and efficacy of drug products. The chapter begins by defining pharmaceutical drying and emphasizing its importance in the manufacturing process. Various methods of pharmaceutical drying, including air drying, vacuum drying, freeze-drying, and spray drying, are discussed, and a comparison between these methods is provided. Factors that influence pharmaceutical drying, such as physical and chemical properties of the product, drying temperature, drying time, pressure, humidity, and solvent properties, are examined. The chapter also highlights the challenges associated with pharmaceutical drying, including product stability and degradation, loss of potency, residual solvents, and the formation of amorphous or crystalline solids. Strategies to overcome these challenges, such as process optimization, the use of drying aids, control of drying parameters, and formulation considerations, are explored. Quality control measures in pharmaceutical drying, including the monitoring of residual moisture and solvent levels, characterization of dried products, and adherence to regulatory guidelines, are discussed.

**Keywords:** pharmaceutical drying, drying methods, factors affecting drying, drying challenges, quality control in drying

## 1. Introduction

Drying is defined as the process of vaporising moisture from a material and wiping it away from the surface, sometimes under vacuum, but probably through carrier gas which passes through or over the material. Drying is popularly conceived as water removal into a hot airstream, but drying could include removing any volatile liquid into any heated gas. Through convection, radiation or conduction, or via internal generation such as dielectric or inductive heating, for drying, as defined, to take place, the moist material must obtain heat from its surroundings; the moisture in the body evaporates and the vapour is received by the carrier gas. In **Figure 1**, this drying process is sketched [1].

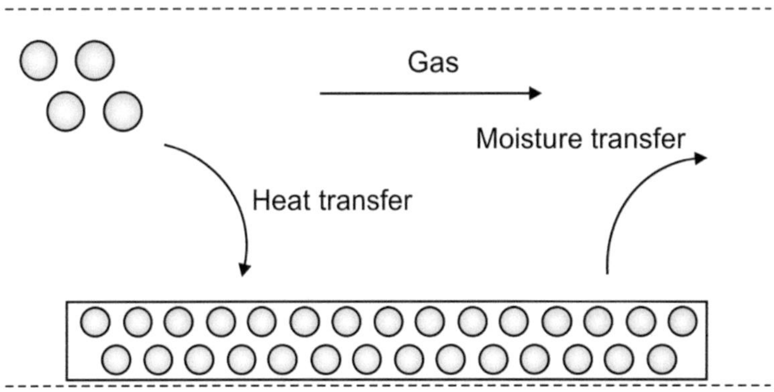


**Figure 1.**
*Drying process.*

Drying has a number of synonyms that are very close. Dehydration is the process where a material is excluded of its water or water is lost as a component. In food-drying operations, the term is often used to describe processes which strive to expel moisture but retain other volatile components in the original material, and that are responsible for valuable aromatic and flavouring properties. A more thorough removal of water is indicated by desiccation. It's being used to indicate almost complete dehydration of these materials for preservation when drying foodstuffs. To describe the thorough removal of moisture from gases, the term is also popularly used. While heat can be used to drive moisture away from a wet substance, by the action of pressure gradients, moisture can be severed with its host material. This process is known as dewatering, and when the moisture-solid bond is not strong, it is generally used as a precursor to the drying of very wet materials. Mechanical means, such as pressing or centrifuging, will be used for dewatering.

Although air is usually considered to be the drying medium, the use of other media does have advantages. If a combustible powder is developed by the solid being dried or the moisture itself is a flammable solvent, then it is advisable to use an inertized or inevitably inert gas. Steam drying has the added advantages of lower use of energy and higher rates of heat transfer. Drying in steam is faster than drying at the same temperature in perfectly dry air above the so-called inversion-point temperature. Moisture is uniformly released during the superheated-steam drying of wood under vacuum. This process is commonly used to produce high-quality seasoned panel timber with minimal degradation caused by drying stresses [2].

For steam drying, a confined vacuum or high-pressure vessel is not necessary. The steam will remain in the dryer by allowing air in the drying chamber to be displaced by water vapour as the vessel warms up and moisture evolution begins, and no complex sealing arrangements are needed for solids to be consumed and released. At 100°C, steam has only 55% of the air density at the same temperature and therefore will remain trapped inside the chamber.

Airless drying is known as the patented method, and the arrangements for batch operation are shown in **Figure 2**. If the vented steam can be used for other purposes, such as hot water production, the airless drying system is capable of showing considerable thermal savings over conventional air drying.

Drying occurs when the wet material contains more moisture than the equilibrium position for its environment. Liquid moisture diffuses to the surface of a wet body where it evaporates, diffusing the vapour into the surrounding air through the

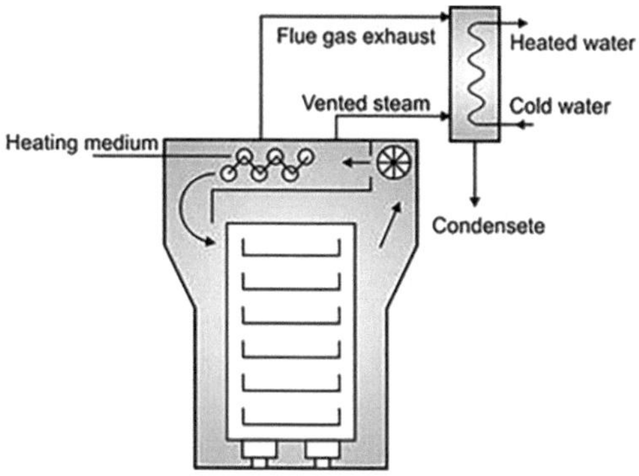


**Figure 2.**
*Airless drying system with heat recovery.*

boundary layer. This was the initial concept of convective drying. This perspective is inadequate, except when drying uniform substances with dissolved moisture [3].

Moisture movement mechanisms are generally more complex. Most materials, such as particles and fibers, are composed of sub-entities that may be loose or held in some kind of matrix. The quantity of moisture retained and the extent of bonding to the solids govern the number and nature of the voids between these entities and the pores within them. The material is said to be capillary-porous if the openings make up a capillary network. A capillary-porous material may be non-hygroscopic: that is, its full vapour pressure is exerted by the moisture held within the body. In some coarse, nonporous mineral aggregates, this is a restrictive case. Between the particles, moisture is simply trapped [4].

### 1.1 Objectives of pharmaceutical drying

The objectives of the drying are;

1. To avoid or eliminate moisture which may lead to corrosion and decrease the product or drug stability.

2. To improve or keep the good properties of a material, e.g. flowability, compressibility.

3. To reduce the cost of transportation of large volume materials (liquids).

4. To make the material easy or more suitable for handling.

5. The final step in evaporation, filtration, crystallization.

6. Preservation of chemical integrity and stability of the dried product.

7. Accelerated drying rates to reduce processing time and energy consumption.

8. Formation of a free-flowing and uniform powder or granular form.

9. Decreasing food decomposition by removing moisture.

10. Preventing decomposition by inhibiting microbial growth.

11. Extending the product's shelf life for a longer duration [5].

## 1.2 Importance of pharmaceutical drying in drug manufacturing

Pharmaceutical drying holds significant importance in drug manufacturing for several reasons [6]:

- *Enhanced stability*: Moisture can lead to degradation of active pharmaceutical ingredients (APIs) and excipients, resulting in reduced efficacy and shelf life. Drying mitigates these concerns and improves the stability of drugs, ensuring their effectiveness throughout the intended shelf life [7].

- *Facilitated processing*: Drying transforms liquid formulations into dry powders or granules, making them easier to handle, package, and transport. It also streamlines downstream processing steps like blending, encapsulation, and tableting [8].

- *Improved solubility*: Certain drugs exhibit better solubility in their dry form. Drying enhances drug solubility, leading to improved bioavailability and therapeutic effectiveness [9].

- *Ensured uniformity*: Drying plays a crucial role in achieving uniformity in drug formulations by eliminating variations in moisture content, which can impact product quality and performance.

- *Increased efficiency*: By reducing the moisture content, drying reduces the weight and volume of drugs, thus enhancing efficiency in transportation, storage, and handling [10].

In this chapter, we will delve deeper into the various methods of pharmaceutical drying, explore the factors influencing the process, address the challenges faced, and discuss strategies for overcoming these challenges. Additionally, quality control measures in pharmaceutical drying, including monitoring residual moisture and solvent levels, characterizing dried products, and adhering to regulatory considerations, will be examined.

## 1.3 Factors affecting pharmaceutical drying

1. *Size uniformity*: Size uniformity refers to the consistency in the size of particles or granules within a sample. It is an important characteristic in various industries such as pharmaceuticals and food processing, as it ensures uniformity in dosage and processing. Size uniformity is typically achieved through proper granulation techniques or sieving processes [11].

2. *Particulate diameter*: Particulate diameter refers to the size of individual particles or granules. It is an essential parameter in various applications such as powder handling, filtration, and particle characterization. Particulate diameter can affect flowability, packing density, dissolution rate, and other physical properties of the material.

3. *Mechanism involved in drying*: Drying mechanisms involve the transfer of moisture from a material to the surrounding environment. Common drying mechanisms include evaporation, diffusion, and convection. Evaporation involves the direct conversion of liquid moisture into vapor, diffusion involves the movement of moisture through the material, and convection utilizes the movement of air or other gases to carry away the evaporated moisture.

4. *Product mass flow rate*: The product mass flow rate refers to the amount of material passing through a specific point in a given time period. It is a measure of the quantity of material being processed or transported in a system. The product mass flow rate is influenced by factors such as the size and design of the equipment, processing parameters, and material properties.

5. *Hot air mass flow rate*: The hot air mass flow rate refers to the amount of heated air or gas used in a drying process. It plays a crucial role in the efficiency and effectiveness of the drying operation. Proper control of the hot air mass flow rate ensures sufficient heat transfer to facilitate moisture evaporation and drying while optimizing energy consumption.

6. *Diameter of the dryer section*: The diameter of the dryer section refers to the size or cross-sectional area of the drying chamber or equipment. It influences the residence time of the material being dried and the drying efficiency. The diameter should be designed to allow for proper airflow and residence time to ensure thorough and efficient drying.

7. *Critical moisture content of the material*: The critical moisture content of a material is the moisture level below which the material becomes stable and resistant to microbial growth, chemical reactions, and degradation. It is an important parameter to determine the endpoint of the drying process and ensure the product's stability and quality during storage.

8. *Physical properties of wet and dry flow particle*: The physical properties of wet and dry flow particles refer to their characteristics in terms of flowability, cohesion, compressibility, and other properties. Wet particles tend to be more cohesive and less flowable due to the presence of moisture, while dry particles exhibit better flowability and reduced cohesion. Understanding these properties is crucial in designing efficient drying processes and handling the dried material effectively.

## 2. Methods of pharmaceutical drying

### 2.1 Air drying

Air drying is a common method used in many industries, including the pharmaceutical industry, to remove moisture from a product or material. It involves the use of ambient air to remove moisture from the surface of the product, typically by evaporation.

The air-drying process involves exposing the material to be dried to a stream of dry, warm air. This air is typically circulated around the material to maximize

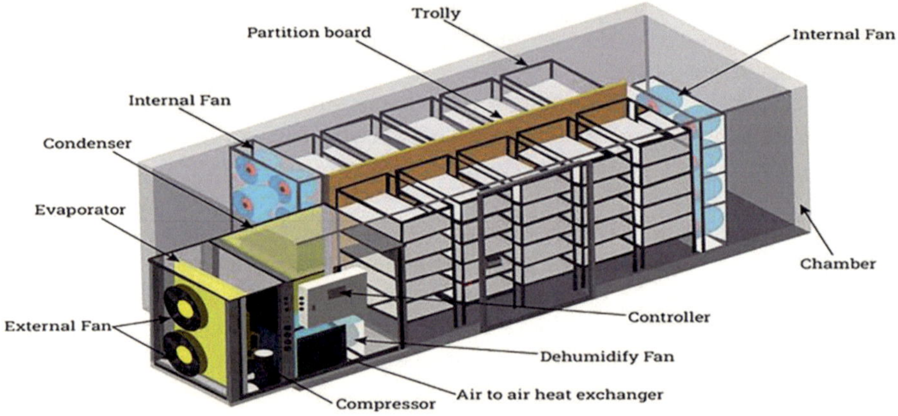


**Figure 3.**
*Industrial air dryer.*

exposure and evaporation. The process is generally slow and can take several hours or even days depending on the size and thickness of the material being dried as shown in **Figure 3** [12].

*2.1.1 Advantages of air-drying method in pharmaceutical industry*

1. *Simple and low-cost method*: Air drying is a simple and cost-effective method of drying in pharmaceutical industry. It requires no specialized equipment, and the only energy required is for circulating the air and maintaining the temperature [13].

2. *Non-destructive*: Air drying is a non-destructive method that can be used for delicate or heat-sensitive materials that cannot be exposed to high temperatures.

3. *Preservation of product quality*: Air drying can help preserve the quality and stability of the product by avoiding exposure to high temperatures that can lead to degradation of active ingredients or changes in physical properties.

4. *Environmentally friendly*: Air drying does not produce any harmful byproducts, making it an environmentally friendly method of drying.

5. *Versatile*: Air drying can be used for a wide range of materials, including heat-sensitive and temperature-stable substances.

*2.1.2 Disadvantages of air-drying method*

1. *Slow process*: Air drying is a slow process and may take longer than other drying methods [14].

2. *Inconsistent results*: The drying rate may vary depending on the humidity and temperature of the environment, which can lead to inconsistent results.

3. *Limited capacity*: Air drying may not be suitable for large-scale production due to its limited capacity.

4. *Risk of contamination*: The exposure of materials to air may increase the risk of contamination by microorganisms or other environmental factors.

*2.1.3 Applications to pharmaceutical industries*

1. *Drying of APIs (active pharmaceutical ingredients)*: Air drying is commonly used for drying APIs, especially those that are sensitive to high temperatures [15].

2. *Drying of excipients*: Excipients are the inactive ingredients used in pharmaceutical formulations. Air drying is used to remove moisture from excipients to prevent degradation or clumping.

3. *Drying of finished dosage forms*: Air drying is used for drying finished dosage forms, such as tablets or capsules, after they are coated with a protective layer.

4. *Drying of raw materials*: Air drying is used to dry raw materials, such as herbs or plant extracts, before they are used in the manufacturing process.

**2.2 Vacuum drying**

Vacuum drying is a method of drying materials by removing moisture under reduced pressure. This method is commonly used in the pharmaceutical industry to remove moisture from heat-sensitive materials, such as APIs and excipients as shown in **Figure 4** [16, 17].

*2.2.1 Advantages of vacuum drying method*

1. *Faster drying*: Vacuum drying is faster than air drying as it uses reduced pressure to remove moisture from materials.

2. *Uniform drying*: Vacuum drying provides a uniform drying rate throughout the material, resulting in consistent and reproducible results.

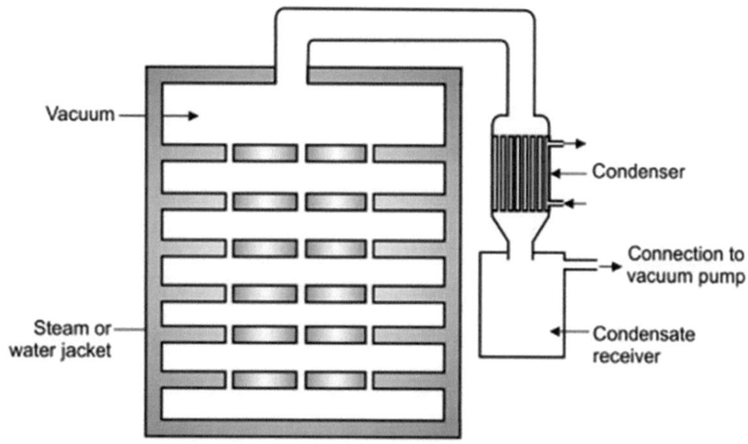


**Figure 4.**
*Vacuum dryer.*

3. *Low temperature*: Vacuum drying operates at low temperatures, which prevents degradation or denaturation of heat-sensitive materials.

4. *Reduced risk of contamination*: Vacuum drying is carried out in a closed system, which reduces the risk of contamination from environmental factors.

5. *Energy-efficient*: Vacuum drying uses less energy than other drying methods as it requires lower temperatures.

#### *2.2.2 Disadvantages of vacuum drying method*

1. *Cost*: Vacuum drying equipment is expensive and may not be feasible for small-scale production.

2. *Maintenance*: Vacuum drying equipment requires regular maintenance to ensure proper operation and prevent contamination.

3. *Complex process*: Vacuum drying is a complex process that requires expertise and specialized equipment.

4. *Limited capacity*: Vacuum drying may not be suitable for large-scale production due to its limited capacity.

#### *2.2.3 Applications to pharmaceutical industries*

1. *Drying of APIs (active pharmaceutical ingredients)*: Vacuum drying is commonly used for drying APIs, especially those that are sensitive to high temperatures [18, 19].

2. *Drying of excipients*: Excipients are the inactive ingredients used in pharmaceutical formulations. Vacuum drying is used to remove moisture from excipients to prevent degradation or clumping.

3. *Drying of finished dosage forms*: Vacuum drying is used for drying finished dosage forms, such as tablets or capsules, after they are coated with a protective layer.

4. *Drying of raw materials*: Vacuum drying is used to dry raw materials, such as herbs or plant extracts, before they are used in the manufacturing process.

### 2.3 Freeze-drying (lyophilization)

Freeze-drying, also known as lyophilization, is a method of drying materials that involves freezing the material and then removing moisture under reduced pressure as shown in **Figure 5**. This method is commonly used in the pharmaceutical industry to preserve the integrity of heat-sensitive materials, such as biologics, vaccines, and other drugs [20].

#### *2.3.1 Advantages of freeze-drying method*

1. *Preserves material integrity*: Freeze-drying preserves the physical and chemical integrity of materials by removing moisture without causing thermal damage.

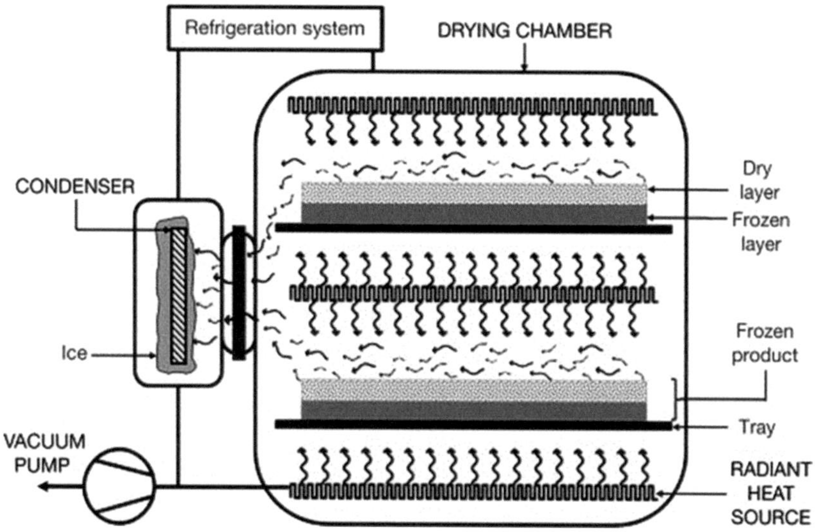


**Figure 5.**
*Freeze-drying process.*

2. *Long shelf-life*: Freeze-dried materials have a longer shelf-life compared to other drying methods, as they are less prone to degradation over time.

3. *Uniform drying*: Freeze-drying provides a uniform drying rate throughout the material, resulting in consistent and reproducible results.

4. *Minimal loss of volatile compounds*: Freeze-drying reduces the loss of volatile compounds, which may be important for certain drug formulations.

5. *Reduced risk of contamination*: Freeze-drying is carried out in a closed system, which reduces the risk of contamination from environmental factors.

### *2.3.2 Disadvantages of freeze-drying method*

1. *Cost*: Freeze-drying equipment is expensive and may not be feasible for small-scale production [21].

2. *Long processing time*: Freeze-drying is a time-consuming process that may take several days or even weeks to complete.

3. *Complex process*: Freeze-drying is a complex process that requires expertise and specialized equipment.

4. *Limited capacity*: Freeze-drying may not be suitable for large-scale production due to its limited capacity.

5. *Requires stability testing*: Freeze-dried materials may require stability testing to ensure that the product retains its physical and chemical properties over time.

### *2.3.3 Applications to pharmaceutical industries*

1. *Drying of biologics and vaccines*: Freeze-drying is commonly used to preserve the integrity of biologics and vaccines, which are often heat-sensitive [22].

2. *Drying of complex drug formulations*: Freeze-drying is used to dry complex drug formulations, such as liposomes and nanoparticles, which may be difficult to dry using other methods.

3. *Preservation of labile drugs*: Freeze-drying is used to preserve the activity of labile drugs, such as enzymes and peptides.

4. *Drying of tissue samples*: Freeze-drying is used to dry tissue samples for long-term storage or analysis.

## 2.4 Spray drying

Spray drying is a widely used method in the pharmaceutical industry for converting a liquid or a solution into a dry powder as shown in **Figure 6**. The process involves atomizing a liquid feed into a spray of small droplets, which are then dried by hot gas streams in a spray dryer. As the droplets travel through the dryer, the solvent evaporates, leaving behind a dry powder [23].

### *2.4.1 Advantages of spray drying method*

1. *High efficiency*: The process of spray drying is a continuous process and can be performed at a high speed with good efficiency.

2. *High yield*: The yield of the product obtained from the spray drying process is high due to the high surface area and efficient drying mechanism.

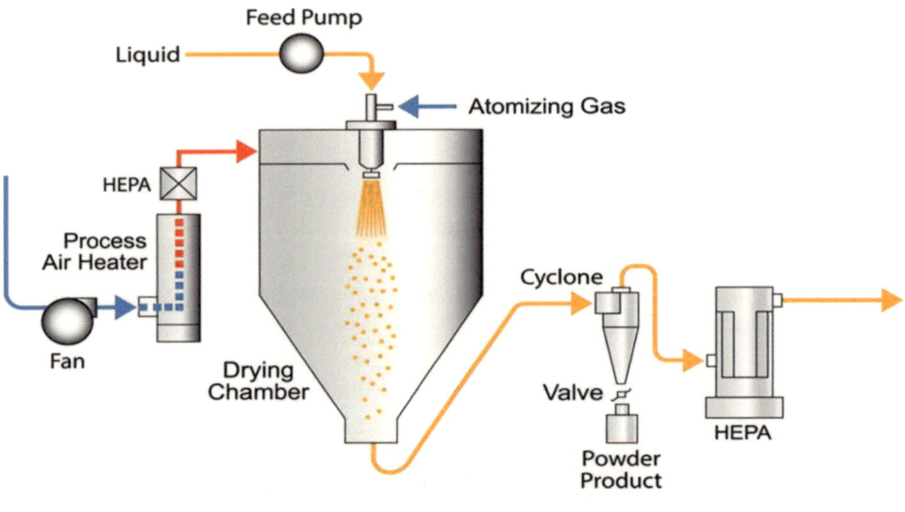


**Figure 6.**
*Spray drying process.*

3. *Better solubility*: Spray drying improves the solubility of the product by reducing the particle size and increasing the surface area, thereby allowing for faster dissolution and absorption of the product.

4. *Preservation of the product*: Spray drying is a gentle method of drying and can be used to preserve the product's physical and chemical properties.

5. *Versatility*: Spray drying can be used to produce a wide range of products, including powders, granules, and agglomerates.

#### 2.4.2 Disadvantages of spray drying method

1. *High cost*: The equipment and maintenance cost of a spray dryer can be quite high.

2. *Heat damage*: Some heat-sensitive products may get damaged or degraded during the spray drying process.

3. *Moisture content*: The spray drying process can leave behind residual moisture in the powder, which can cause stability issues.

4. *Particle size*: The particle size distribution of the powder obtained from the spray drying process can be broad, which can cause issues in downstream processes.

#### 2.4.3 Applications of spray drying method in pharmaceutical industries

1. *Inhalable powders*: The spray drying method is widely used to produce inhalable powders for pulmonary drug delivery [24].

2. *Oral solid dosage forms*: Spray drying can be used to produce solid dosage forms such as tablets and capsules by using spray-dried granules or directly compressing the spray-dried powder.

3. *Parenteral formulations*: The spray drying method can be used to produce parenteral formulations such as injectables and lyophilized products.

4. *Taste-masking*: Spray drying can be used to improve the taste of bitter or unpleasant-tasting drugs by incorporating them into a taste-masking matrix.

### 2.5 Comparison of different drying methods

There are several different methods for drying materials, including air drying, freeze-drying, spray drying, and vacuum drying. Each method has its advantages and disadvantages, and the choice of drying method depends on the nature of the material being dried and the desired characteristics of the final product [25–29].

1. *Air drying*: Air drying is a simple and low-cost method of drying that involves exposing the material to air and allowing it to dry naturally. This method is suitable for drying materials that are not sensitive to heat and that do not require rapid drying. However, air drying can be slow and may not be suitable for materials that require precise control of temperature and humidity.

2. *Freeze drying*: Freeze-drying, also known as lyophilization, is a method of drying that involves freezing the material and then removing the ice by sublimation under vacuum. This method is suitable for drying materials that are sensitive to heat and that require preservation of their biological activity. Freeze-drying can produce a high-quality product with good shelf stability, but it can be a slow and expensive process.

3. *Spray drying*: Spray drying is a method of drying that involves atomizing a liquid or solution into a spray of small droplets, which are then dried by hot gas streams in a spray dryer. This method is suitable for producing dry powders with good solubility and dispersibility. Spray drying is a fast and efficient method of drying, but it can be expensive and may not be suitable for heat-sensitive materials.

4. *Vacuum drying*: Vacuum drying is a method of drying that involves applying heat and reduced pressure to the material. This method is suitable for drying materials that are sensitive to oxidation and that require precise control of temperature and humidity. Vacuum drying can produce a high-quality product, but it can be a slow and expensive process.

In summary, each drying method has its advantages and disadvantages, and the choice of method depends on the nature of the material being dried and the desired characteristics of the final product. Air drying is a simple and low-cost method, freeze-drying is suitable for preserving biological activity, spray drying is suitable for producing dry powders with good solubility and dispersibility, and vacuum drying is suitable for heat-sensitive materials that require precise control of temperature and humidity.

## 3. Challenges in pharmaceutical drying

Pharmaceutical drying is a critical process in the manufacturing of drugs and other pharmaceutical products. It involves the removal of moisture from the products to ensure stability, potency, and efficacy. However, there are several challenges associated with pharmaceutical drying that can affect the quality of the final product. Here are the challenges in detail [30–33].

### 3.1 Product stability and degradation

Pharmaceutical products are sensitive to heat and moisture, and the drying process can cause degradation, which can lead to changes in the chemical and physical properties of the product. The stability and shelf-life of a drug depend on its formulation and the conditions under which it is stored. Heat and moisture can cause changes in the drug's molecular structure, leading to loss of potency and reduced efficacy. Therefore, it is crucial to optimize the drying conditions to avoid excessive exposure to heat and moisture.

### 3.2 Loss of potency

Some drugs are sensitive to heat, and the drying process can cause a loss of potency. The drying conditions must be carefully optimized to ensure that the drugs are dried efficiently without compromising their potency. The drying temperature, time, and humidity must be carefully monitored to prevent excessive exposure of the drug to heat and moisture.

### 3.3 Residual solvents

Some pharmaceutical products require solvents to dissolve the active ingredients. Residual solvents can be left behind after the drying process, and these can be harmful to the patient. Therefore, it is essential to use proper solvent removal techniques to ensure that no residual solvent is left behind in the final product.

### 3.4 Formation of amorphous or crystalline solids

The drying process can cause the formation of amorphous or crystalline solids, which can affect the drug's solubility, stability, and bioavailability. Amorphous solids are less stable than crystalline solids, and they tend to have a shorter shelf-life. On the other hand, crystalline solids can be challenging to dissolve, leading to reduced bioavailability. Therefore, it is essential to optimize the drying conditions to control the physical form of the final product.

The challenges associated with pharmaceutical drying can affect the quality, stability, and efficacy of the final product. Therefore, it is crucial to optimize the drying conditions to ensure that the products are dried efficiently without compromising their stability and potency. The drying process must be carefully monitored to prevent the formation of amorphous or crystalline solids, and proper solvent removal techniques must be employed to prevent the presence of residual solvents.

## 4. Strategies for overcoming drying challenges

Drying is an essential step in the manufacturing of pharmaceuticals, and it is often associated with several challenges that can affect the quality, stability, and efficacy of the final product. However, there are several strategies that can be employed to overcome these drying challenges. Here are some of the strategies in detail [34–38].

### 4.1 Process optimization

Process optimization involves optimizing the drying conditions to achieve the desired drying rate and ensure the stability, potency, and efficacy of the final product. Process optimization includes selecting the appropriate drying method, adjusting the drying parameters such as temperature, humidity, airflow rate, and time, and choosing the appropriate equipment for the drying process. The goal is to ensure that the drying process is efficient and effective, while minimizing any adverse effects on the product.

### 4.2 Use of drying aids

Drying aids are substances that are added to the product during the drying process to improve the drying efficiency and prevent degradation. Examples of drying aids include desiccants, which absorb moisture from the product, and inert gases such as nitrogen, which can help to prevent oxidation during the drying process. Drying aids can also be used to control the physical form of the final product, such as the use of surfactants to control the particle size distribution.

### 4.3 Control of drying parameters

The control of drying parameters is crucial to ensure the stability and efficacy of the final product. The drying parameters that need to be controlled include temperature, humidity, airflow rate, and time. It is essential to monitor and control these parameters to prevent over-drying or under-drying, which can lead to product degradation or loss of potency. Advanced process control techniques such as feedback control systems can be used to control the drying parameters automatically.

### 4.4 Formulation considerations

Formulation considerations involve selecting the appropriate formulation for the product to optimize the drying process. The formulation can affect the drying rate, the physical form of the final product, and the stability and efficacy of the drug. Formulation considerations include the selection of the appropriate excipients to stabilize the drug, the optimization of the particle size distribution to ensure efficient drying, and the use of amorphous or crystalline forms to optimize solubility and stability.

Overcoming drying challenges in pharmaceutical manufacturing requires a multifaceted approach that involves process optimization, the use of drying aids, the control of drying parameters, and formulation considerations. It is essential to optimize the drying process to ensure the stability, potency, and efficacy of the final product, while minimizing any adverse effects on the product. By employing these strategies, it is possible to overcome the challenges associated with drying and produce high-quality pharmaceutical products.

## 5. Quality control in pharmaceutical drying

Quality control in pharmaceutical drying is an essential aspect of the manufacturing process, and it involves various measures to ensure the final product's quality, safety, and efficacy. Here are some of the quality control measures in detail [39, 40].

### 5.1 Monitoring of residual moisture and solvent levels

Monitoring of residual moisture and solvent levels is a critical aspect of quality control in pharmaceutical drying. The residual moisture and solvent levels in the final product can affect its stability, safety, and efficacy. To ensure the product quality, manufacturers need to monitor and control these parameters throughout the drying process.

The residual moisture content is the amount of moisture remaining in the dried product after the drying process is complete. The residual moisture content can affect the stability and shelf-life of the final product. If the residual moisture content is too high, it can promote microbial growth, chemical reactions, and degradation. If the residual moisture content is too low, the product may become brittle or hard.

The solvent levels in the final product need to be monitored to prevent toxicity and adverse effects. The solvents used in the drying process may be toxic, and their residual levels need to be within the acceptable limits. The regulatory authorities specify the maximum residual levels of solvents in the final product.

Various analytical methods are available to monitor the residual moisture and solvent levels. Karl Fischer titration is a widely used method to determine the residual moisture content. The method involves titrating the sample with a Karl Fischer reagent, which reacts with the water present in the sample. The amount of reagent consumed is proportional to the amount of water in the sample, which is used to calculate the residual moisture content.

Gas chromatography and high-performance liquid chromatography (HPLC) are commonly used methods to determine residual solvent levels. These methods involve separating the solvent from the sample and analyzing its concentration.

### 5.2 Characterization of dried products

Characterization of dried products is an important aspect of quality control in pharmaceutical drying. It involves evaluating the physical, chemical, and structural properties of the dried product to ensure its quality, safety, and efficacy. The characterization process is done using various analytical techniques, and the results are compared with the specifications to ensure compliance.

Physical characterization of the dried product includes measuring the size, shape, density, and porosity of the particles. These properties affect the performance and processing characteristics of the product. For example, the particle size distribution can affect the flowability, solubility, and bioavailability of the product. The density and porosity can affect the compressibility and dissolution rate of the product. Physical properties can be characterized using techniques such as particle size analysis, microscopy, surface area analysis, and compressibility testing.

Chemical characterization of the dried product includes evaluating the purity, identity, and stability of the product. These properties affect the safety and efficacy of the product. For example, the purity of the product ensures that it does not contain impurities that can cause adverse effects. The identity of the product ensures that it is the desired compound and not a different compound or an isomer. The stability of the product ensures that it remains effective and safe throughout its shelf life. Chemical properties can be characterized using techniques such as chromatography, spectroscopy, and thermal analysis.

Structural characterization of the dried product includes determining the crystalline or amorphous nature of the product. This property affects the dissolution and bioavailability of the product. For example, the amorphous form of the product has a higher dissolution rate and bioavailability than the crystalline form. Structural properties can be characterized using techniques such as X-ray diffraction, differential scanning calorimetry, and solid-state nuclear magnetic resonance.

Regulatory authorities such as the United States Food and Drug Administration (FDA) require pharmaceutical manufacturers to demonstrate that their products meet the specifications outlined in the drug application. Therefore, characterization of dried products is an essential part of the quality control process in pharmaceutical manufacturing. It ensures that the product is safe, effective, and consistent throughout its shelf life.

### 5.3 Regulatory considerations

Regulatory considerations are an essential aspect of quality control in pharmaceutical drying. The manufacturing process needs to comply with various regulations and guidelines such as Good Manufacturing Practices (GMP) and the International Conference on Harmonization (ICH) guidelines. The regulatory authorities require the manufacturers to demonstrate the safety, efficacy, and quality of the final product through various tests and analyses. The manufacturers need to provide documentation of the manufacturing process and the quality control measures employed.

## 6. Conclusions

Pharmaceutical drying is a critical step in drug manufacturing, aiming to remove moisture and maintain product stability and efficacy.

Different drying methods like air drying, vacuum drying, freeze-drying, and spray drying are used based on specific product requirements.

The efficiency of pharmaceutical drying is influenced by factors such as physical and chemical properties, temperature, time, pressure, humidity, and solvent properties.

Challenges in pharmaceutical drying include product stability, potency loss, residual solvents, and formation of amorphous or crystalline solids.

Strategies for overcoming these challenges involve process optimization, the use of drying aids, control of drying parameters, and formulation considerations.

Quality control measures including monitoring residual moisture and solvent levels, characterizing dried products, and adhering to regulatory standards are crucial for ensuring the final product's quality and safety.

## Acknowledgements

The authors would like to express their sincere gratitude to the Management of RSM's N.N. Sattha College of Pharmacy, Ahmednagar, for providing the necessary resources and support for the completion of this work. We would also like to acknowledge the valuable contributions of our colleagues who provided insightful discussions and suggestions throughout the research process.

## Conflict of interest

The authors declare no conflict of interest.

## Author details

Sachin Kothawade*, Vishal Pande, Vaibhav Wagh, Kalyani Autade, Sandesh Bole, Rajashri Sumbe and Shubhangi Albhar
RSM's N.N. Sattha College of Pharmacy, Ahmednagar, MH, India

*Address all correspondence to: sachin.kothawade23@gmail.com

## References

[1] Bhujbal SV, Mitra B, Jain U, Gong Y, Agrawal A, Karki S, et al. Pharmaceutical amorphous solid dispersion: A review of manufacturing strategies. Acta Pharmaceutica Sinica B. 2021;**11**(8):2505-2536. DOI: 10.1016/j.apsb.2021.05.014

[2] Rezvankhah A, Emam-Djomeh Z, Askari G. Encapsulation and delivery of bioactive compounds using spray and freeze-drying techniques: A review. Drying Technology. 2020;**38**(1-2):235-258. DOI: 10.1080/07373937.2019.1653906

[3] Anwar SH, Kunz B. The influence of drying methods on the stabilization of fish oil microcapsules: Comparison of spray granulation, spray drying, and freeze drying. Journal of Food Engineering. 2011;**105**(2):367-378. DOI: 10.1016/j.jfoodeng.2011.02.047

[4] Thoorens G, Krier F, Leclercq B, Carlin B, Evrard B. Microcrystalline cellulose, a direct compression binder in a quality by design environment—A review. International Journal of Pharmaceutics. 2014;**473**(1-2):64-72. DOI: 10.1016/j.ijpharm.2014.06.055

[5] Farah S, Anderson DG, Langer R. Physical and mechanical properties of PLA, and their functions in widespread applications—A comprehensive review. Advanced Drug Delivery Reviews. 2016;**107**:367-392. DOI: 10.1016/j.addr.2016.06.012

[6] Langford A, Bhatnagar B, Walters R, Tchessalov S, Ohtake S. Drying technologies for biopharmaceutical applications: Recent developments and future direction. Drying Technology. 2018;**36**(6):677-684. DOI: 10.1080/07373937.2017.1355318

[7] Mašková E, Kubová K, Raimi-Abraham BT, Vllasaliu D, Vohlídalová E, Turánek J, et al. Hypromellose–A traditional pharmaceutical excipient with modern applications in oral and oromucosal drug delivery. Journal of Controlled Release. 2020;**324**:695-727. DOI: 10.1016/j.jconrel.2020.05.045

[8] Darji MA, Lalge RM, Marathe SP, Mulay TD, Fatima T, Alshammari A, et al. Excipient stability in oral solid dosage forms: A review. AAPS PharmSciTech. 2018;**19**:12-26. DOI: 10.1208/s12249-017-0864-4

[9] Szabó E, Démuth B, Galata DL, Vass P, Hirsch E, Csontos I, et al. Continuous formulation approaches of amorphous solid dispersions: Significance of powder flow properties and feeding performance. Pharmaceutics. 2019;**11**(12):654. DOI: 10.3390/pharmaceutics11120654

[10] Khadka P, Ro J, Kim H, Kim I, Kim JT, Kim H, et al. Pharmaceutical particle technologies: An approach to improve drug solubility, dissolution and bioavailability. Asian Journal of Pharmaceutical Sciences. 2014;**9**(6):304-316. DOI: 10.1016/j.ajps.2014.05.005

[11] Babu AK, Kumaresan G, Raj VA, Velraj R. Review of leaf drying: Mechanism and influencing parameters, drying methods, nutrient preservation, and mathematical models. Renewable and Sustainable Energy Reviews. 2018;**90**:536-556. DOI: 10.1016/j.rser.2018.04.002

[12] Pawar SB, Pratape VM. Fundamentals of infrared heating and its application in drying of food materials: A review. Journal of Food Process Engineering. 2017;**40**(1):e12308. DOI: 10.1111/jfpe.12308

[13] Ruiz-Medina MA, Sansón M, González-Rodríguez ÁM. Changes in antioxidant activity of fresh marine macroalgae from the Canary Islands during air-drying process. Algal Research. 2022;**66**:102798. DOI: 10.1016/j.algal.2022.102798

[14] Dehghannya J, Seyed-Tabatabaei SR, Khakbaz Heshmati M, Ghanbarzadeh B. Influence of three stage ultrasound—intermittent microwave—hot air drying of carrot on physical properties and energy consumption. Heat and Mass Transfer. 2021;**57**(12):1893-1907. DOI: 10.1007/s00231-021-03074-1

[15] Ohtake S, Izutsu KI, Lechuga-Ballesteros D, editors. Drying Technologies for Biotechnology and Pharmaceutical Applications. New Jersy, USA: John Wiley & Sons; 2020. DOI: 10.1002/9783527802104

[16] McLoughlin CM, McMinn WA, Magee TR. Microwave-vacuum drying of pharmaceutical powders. Drying Technology. 2003;**21**(9):1719-1733. DOI: 10.1081/DRT-120025505

[17] Abano EE, Ma H, Qu W. Influence of combined microwave-vacuum drying on drying kinetics and quality of dried tomato slices. Journal of Food Quality. 2012;**35**(3):159-168. DOI: 10.1111/j.1745-4557.2012.00446.x

[18] Walters RH, Bhatnagar B, Tchessalov S, Izutsu KI, Tsumoto K, Ohtake S. Next generation drying technologies for pharmaceutical applications. Journal of Pharmaceutical Sciences. 2014;**103**(9):2673-2695. DOI: 10.1002/jps.23998

[19] Farrel G, McMinn WA, Magee TR. Microwave-vacuum drying kinetics of pharmaceutical powders. Drying Technology. 2005;**23**(9-11):2131-2146. DOI: 10.1080/07373930500212354

[20] Sadikoglu H, Ozdemir M, Seker M. Freeze-drying of pharmaceutical products: Research and development needs. Drying Technology. 2006;**24**(7):849-861. DOI: 10.1080/07373930600734018

[21] Pisano R, Fissore D, Barresi AA, Rastelli M. Quality by design: Scale-up of freeze-drying cycles in pharmaceutical industry. AAPS PharmSciTech. 2013;**14**:1137-1149. DOI: 10.1208/s12249-013-0003-9

[22] Wanning S, Süverkrüp R, Lamprecht A. Pharmaceutical spray freeze drying. International Journal of Pharmaceutics. 2015;**488**(1-2):136-153. DOI: 10.1016/j.ijpharm.2015.04.053

[23] Samborska K, Poozesh S, Barańska A, Sobulska M, Jedlińska A, Arpagaus C, et al. Innovations in spray drying process for food and pharma industries. Journal of Food Engineering. 2022;**2022**:110960. DOI: 10.1016/j.jfoodeng.2022.110960

[24] Sollohub K, Cal K. Spray drying technique: II. Current applications in pharmaceutical technology. Journal of Pharmaceutical Sciences. 2010;**99**(2):587-597. DOI: 10.1002/jps.21963

[25] Si X, Chen Q, Bi J, Wu X, Yi J, Zhou L, et al. Comparison of different drying methods on the physical properties, bioactive compounds and antioxidant activity of raspberry powders. Journal of the Science of Food and Agriculture. 2016;**96**(6):2055-2062. DOI: 10.1002/jsfa.7317

[26] Farahmandfar R, Tirgarian B, Dehghan B, Nemati A. Comparison of different drying methods on bitter orange (*Citrus aurantium* L.) peel waste: Changes in physical (density and color) and essential oil (yield, composition, antioxidant and antibacterial)

properties of powders. Journal of Food Measurement and Characterization. 2020;**14**:862-875. DOI: 10.1007/s11694-019-00334-x

[27] An K, Zhao D, Wang Z, Wu J, Xu Y, Xiao G. Comparison of different drying methods on Chinese ginger (*Zingiber officinale* Roscoe): Changes in volatiles, chemical profile, antioxidant properties, and microstructure. Food Chemistry. 2016;**197**:1292-1300. DOI: 10.1016/j.foodchem.2015.11.033

[28] Abdul-Fattah AM, Kalonia DS, Pikal MJ. The challenge of drying method selection for protein pharmaceuticals: Product quality implications. Journal of Pharmaceutical Sciences. 2007;**96**(8):1886-1916. DOI: 10.1002/jps.20842

[29] Kaialy W. A review of factors affecting electrostatic charging of pharmaceuticals and adhesive mixtures for inhalation. International Journal of Pharmaceutics. 2016;**503**(1-2):262-276. DOI: 10.1016/j.ijpharm.2016.01.076

[30] Chakraborty J, Subash M, Thorat BN. Drying induced polymorphic transformation of pharmaceutical ingredients: A critical review of recent progresses and challenges. Drying Technology. 2022;**40**(14):2817-2835. DOI: 10.1080/07373937.2021.1983823

[31] Conder EW, Cosbie AS, Gaertner J, Hicks W, Huggins S, MacLeod CS, et al. The pharmaceutical drying unit operation: An industry perspective on advancing the science and development approach for scale-up and technology transfer. Organic Process Research and Development. 2017;**21**(3):420-429. DOI: 10.1021/acs.oprd.6b00406

[32] Vass P, Démuth B, Hirsch E, Nagy B, Andersen SK, Vigh T, et al. Drying technology strategies for colon-targeted oral delivery of biopharmaceuticals. Journal of Controlled Release. 2019;**296**:162-178. DOI: 10.1016/j.jconrel.2019.01.023

[33] Cote A, Erdemir D, Girard KP, Green DA, Lovette MA, Sirota E, et al. Perspectives on the current state, challenges, and opportunities in pharmaceutical crystallization process development. Crystal Growth & Design. 2020;**20**(12):7568-7581. DOI: 10.1021/acs.cgd.0c00847

[34] De Kesel PM, Sadones N, Capiau S, Lambert WE, Stove CP. Hemato-critical issues in quantitative analysis of dried blood spots: Challenges and solutions. Bioanalysis. 2013;**5**(16):2023-2041. DOI: 10.4155/bio.13.156

[35] Sosnik A, Seremeta KP. Advantages and challenges of the spray-drying technology for the production of pure drug particles and drug-loaded polymeric carriers. Advances in Colloid and Interface Science. 2015;**223**:40-54. DOI: 10.1016/j.cis.2015.05.003

[36] Sosnik A, Augustine R. Challenges in oral drug delivery of antiretrovirals and the innovative strategies to overcome them. Advanced Drug Delivery Reviews. 2016;**103**:105-120. DOI: 10.1016/j.addr.2015.12.022

[37] Chang RY, Chen L, Chen D, Chan HK. Overcoming challenges for development of amorphous powders for inhalation. Expert Opinion on Drug Delivery. 2020;**17**(11):1583-1595. DOI: 10.1080/17425247.2020.1813105

[38] d'Angelo I, Conte C, La Rotonda MI, Miro A, Quaglia F, Ungaro F. Improving the efficacy of inhaled drugs in cystic fibrosis: challenges and emerging drug delivery strategies. Advanced Drug Delivery Reviews. 2014;**75**:92-111. DOI: 10.1016/j.addr.2014.05.008

[39] Destro F, Barolo M, Nagy ZK. Quality-by-control of intensified continuous filtration-drying of active pharmaceutical ingredients. AICHE Journal. 2023;**69**(2):e17926. DOI: 10.1002/aic.17926

[40] Andrieu J, Vessot S. A review on experimental determination and optimization of physical quality factors during pharmaceuticals freeze-drying cycles. Drying Technology. 2018;**36**(2):129-145. DOI: 10.1080/07373937.2017.1340895

Chapter 6

# Excipient-Free Spray Drying of Bioactive Recombinant Proteins Produced in Plants

*Natalía Reynisdóttir, Páll Thor Ingvarsson, Ásta María Einarsdóttir, Arnór Freyr Ingunnarson and Ildikó Nagy*

## Abstract

Spray drying is an economical drying method for converting aqueous solutions into stable dry powders. This one-step continuous process generates a sustainable solution for long-term storage of various protein formulations. This study focuses on recombinant growth factors produced in a barley seed host. The retained bioactivity of the growth factor in the final solid form suggests that co-purified host components may have preserving effects throughout the optimized spray drying process. To identify the critical spray drying parameters, a customized response surface design of experiment was applied. The defined input spray drying parameters: feed flow rate, spray gas flow rate, and outlet temperature, as well as their interactions, were discovered to be the most critical in terms of product quality and yield. The best operating parameters were chosen after considering potential reduction in energy consumption of the process. Cell proliferation assay results, which show the bioactivity of the growth factors, reveal that the native host components seem to act as proper stabilizing agents that protect the fragile growth factors against various stresses during the drying procedure. This unique matrix composition therefore surpasses the time-consuming process optimization with excipients, allowing for a fully continuous process from purification to the final formulated powder.

**Keywords:** spray drying, design of experiment, optimization, preservation matrix, bioactive growth factors, barley proteins, economical drying, continuous system

## 1. Introduction

Plant molecular farming has grown and advanced immensely into a viable platform to produce commercial proteins [1]. Plant systems have many advantages over their counterparts, mostly in terms of ease in process scale-up, but they are also cost-effective, versatile, and robust. After harvest, plant hosts like the barley seed can be stockpiled under ambient conditions for many years, which largely benefit protein production and enable separation of downstream from upstream processing. Crude materials can therefore be collected and stored long term, and purification can occur at convenience, reinforcing the flexibility of the system [2].

Expression hosts for the production of recombinant proteins, like animal growth factors, mostly range from prokaryotic to eukaryotic systems, such as bacteria, yeast, insects, and mammalian cell cultures. The reason being, these already established expression host systems have been well-defined with current good manufacturing practice (cGMP) and generally offer high production capacity at low cost [3]. Transgenic plants, however, are more unconventional but have in the past two decades emerged in biopharma as an alternative production system [4]. One of those innovative plant systems is the barley plant, which displays a biologically contained expression host that is convenient in various ways. Barley has a generally regarded as safe (GRAS) status and constitutes an endotoxin-free host expression system, as it avoids bacterial host cell-derived endotoxins [5]. Also, barley does not contain secondary metabolites or mammalian-derived pathogens, such as virus infections [6].

The downstream processing involves extracting and purifying the recombinant growth factors from the barley seed, resulting in an aqueous protein solution. The recombinant growth factor is co-purified along with beneficial barley components thought to enhance the final product, as well as ease the purification procedure. The inherent instability of proteins in aqueous solutions is caused by the molecular mobility in solutions. This has been overcome by, for example, storing and transporting the proteins under frozen conditions [7], but dry powder formulations of protein simplify all handling, storage, and distribution, offering an intriguing alternative [7]. The final formulation of purified growth factors is therefore more conveniently given as dried powder with preserved bioactivity for effective storage, easier handling, and more economic shipping options. Dried formulations can circumvent the need for cold chain transport, thereby offering an economically feasible solution for storing and transporting more growth factor per weight at ambient conditions [8].

Spray drying transforms liquid feed into dry particles. However, unlike freeze drying, spray drying is a continuous process that dries the product material in a single manufacturing step. During the spray drying of proteins, the feed solution is atomized into heated air, and the solvent evaporates quickly, leaving behind dry particles that need to be separated from the airstream and collected [9]. This rapid solidification prevents the molecules from arranging into crystal lattices, allowing mainly just amorphous particles to form. Homogenous powders are produced by the spray drying process as a result [10]. This process is illustrated by Büchi (*Büchi Labortechnik AG, Switzerland)* in **Figure 1**.

Some of the process parameters that can be optimized in spray drying include the inlet/outlet temperature, spray gas flow rate (atomizing gas), drying gas flow rate (aspirator rate), liquid feed flow rate (FFR), and concentration and composition of the solution [12]. The outlet temperature is usually considered as a dependent variable, resulting from the combination of the other parameters, although the droplet/particle temperature will never exceed the value of the outlet temperature [13], making it a critical parameter on its own for thermosensitive molecules, such as growth factors. The process is optimized by testing different parameter combinations. The resulting product quality is evaluated based on particle characteristics, including size, shape, flowability, density, and moisture content. All these output effects depend on the selected combination of the process parameters, the solvent, and whether any excipients have been added.

Although freeze drying has been recognized as the gold standard of drying methods [14], traditionally being considered the process of choice for improving the

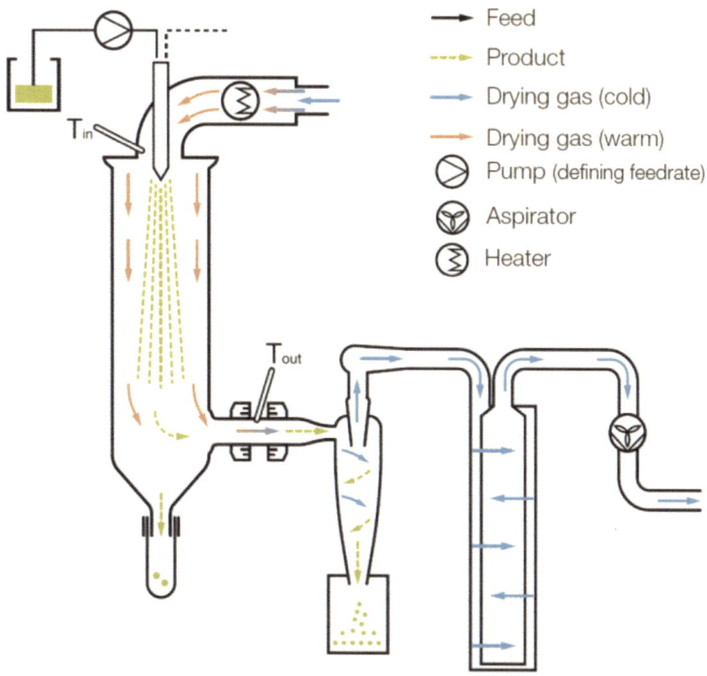


**Figure 1.**
*Spray drying functional principle. With permission from Büchi Labortechnik AG, Switzerland [11].*

long-term storage of manufactured protein, there are some essential challenges to consider. Among others, those are the process scalability, batch-to-batch variance, occurrence of crystallization, and high associated costs [15]. Freeze drying is also restricted by an economic drawback that relates to long and energy-consuming processing time [16]. Consequently, spray drying has started gaining more prevalence as a reliable drying formulation method. That includes protein formulations in established and highly regulated industries such as biopharma [17]. Even though spray drying is considered to be a gentle drying process [17], it is intrinsically more aggressive than freeze drying, since the product is introduced to hot gas during evaporation of the droplets that are sprayed during the process. Therefore, the spray drying process still needs to be carefully examined, especially for the more sensitive proteins, like growth factors.

Drying and dehydration, in general, whether it is spray or freeze drying, puts additional stress on proteins. During drying, hydrogen bonds supplied by water are broken, which may cause conformational changes and thereby inactivity of the protein. The process can expose the product to various interface and shear stresses, which could result in reduced product stability during storage [8, 18, 19]. To counteract this, excipients may be introduced into the formulations. These excipients must replace the hydrogen bonds formerly supplied by water (water replacement theory) and form a viscous matrix around the protein molecules to hinder any molecular motion (vitrification theory), preferably with a high glass transition temperature to increase the storage stability [15, 17]. Traditionally, these excipients have been *i)* non-reducing sugars, such as trehalose and sucrose [15]; *ii)* sugar alcohols such as mannitol and sorbitol [15]; *iii)* oligo- and polysaccharides, such as dextrins and dextrans [15]; and *iv)* single amino acid, such as arginine, leucine, and glycine [15]. For example, disaccharides are excipients thought to have a stabilizing effect on proteins by forming direct hydrogen bonding with them in solid state, thus stabilizing the structure, in addition to forming a highly viscous matrix around the proteins, slowing down molecular movements and thereby degradation [20, 21]. They assist in maintaining the homogeneous and native protein structure, resulting in a stable formulation [20, 22, 23]. Additionally, surfactants may be required to eliminate protein surface adsorption to the abundant air liquid interface created in the atomization step [15, 17].

*Pinto et al.* [15] recently published a comprehensive discussion on the latest development of modern-day trends and the contemporary progress in protein pharmaceutical formulation. Matrix forming excipients are commonly used for protecting protein molecules during the spray drying process and storage [15]. Knowledge about formulation and selection of excipients for freeze-drying proteins can often be applied for spray drying as well [24]. Although excipients are widely recognized for effectively stabilizing many protein solids for freeze-drying formulation [25], *Chen et al.* [26] recently pointed out that a systematic examination of excipient effects on protein stability in spray-dried solids, specifically, is still limited. Critical understanding of the respective interrelation of protein-matrix is still lacking, in addition to a deficient understanding of the storage stability of spray-dried material [26].

For spray drying of proteins, other proteins, such as various albumins, have been studied as excipients in spray-dried formulations [15]. They have even been shown to competitively occupy the particle surface, thereby protecting the protein of interest against surface accumulation and concomitant deactivation [27]. Barley proteins are interesting excipients for the food industry and are considered valued external excipients for the encapsulation of some bioactive ingredients both in spray drying [28] and in freeze drying [29]. It was shown by *Wang et al.* that encapsulation of fish oil with barley proteins had a protective effect against oxidation [28] and by *Meira et al.* that barley residue proteins from beer waste could be used as coating material in microencapsulation of β-carotene [29]. This makes the use of background barley proteins for growth factor dry powder stabilization interesting to investigate further.

Limited published work exists for the spray drying of growth factors explicitly. Growth factors are bioactive proteins that stimulate cell proliferation and differentiation. They have a collective function to expand, maintain, and differentiate cells. There is precedent for spray drying basic fibroblast growth factor (FGF-b) and insulin-like growth factor 1 (IGF-1). In these studies [30, 31], the growth factors are expressed in different host systems than barley. Industry-wide, FGF-b is notorious for being problematic to work with and is constrained by its lack of stability, especially in aqueous solutions. Due to its rapid degradation rate, formulating FGF-b into a reliable product has remained a great challenge [32]. *Ibrahim et al.* [30] developed a spray-dried FGF-b using lactose and leucine as excipients, among others, lactose being a well-known matrix former [15] and leucine producing a hydrophobic surface due to its surfactant properties. IGF-1 promotes cell growth by resulting in a higher cell density and reducing cell death [33]. *Schultz et al.* [31] showed for IGF-1, encapsulated in trehalose, that the bioactivity remained unaffected after spray drying. In terms of recombinant proteins, a recent study by *Vilatte et al.* [34] illustrates spray drying as a viable preservation technology for recombinant proteins produced in microalgae.

Current study aims to investigate the suitability of spray drying recombinant growth factors generated in the barley seed host, co-purified with other host barley components. Generally, recombinant growth factors are fully purified to exclude other components derived from the expression system, but here, other barley components are still present during the preparation of the final product. The goal of this research is to provide usable findings to other scientists working with plant expression systems to produce recombinant proteins. This is the only existing case study that covers the spray drying of recombinant protein expressed in barley. To the best of our knowledge, co-purifying barley has no precedence.

## 2. Case study: spray drying optimization

The following case study investigated whether an economical spray drying procedure could be developed with a wide range of operating conditions, where the bioactivity of the growth factor would be preserved when the target protein is embedded in native barley background and the powder quality could be optimized.

### 2.1 Recombinant growth factors at ORF Genetics

Barley plant (*Hordeum vulgare*) is used at ORF Genetics for the production of recombinant growth factors. The barley grain has a natural inert storage environment to preserve proteins and nutrients for the growing embryo [5]. At large, the nutritional profile of barley consists of starch (65%–68%), total protein (10%–17%), free lipids (2%–3%), β-glucans (4%–9%), and minerals (1.5%–2.5%) [35].

The recombinant growth factor is expressed in the endosperm tissue of the barley seed. The production expression system has been further developed and optimized at ORF Genetics, Iceland. After harvesting, dehulling, and milling of the seeds, the target protein is extracted along with the native barley proteins in an aqueous buffer solution. The suspension is then centrifuged, further purified, and concentrated. The semi-purified growth factor solution then undergoes buffer exchange in preparation for the final formulation.

The continuously expanding MESOkine® portfolio at ORF Genetics represents high-quality, plant-made, endotoxin-free animal recombinant growth factors available for the cell-cultured meat (CCM) industry (https://www.orfgenetics.com/). However, for this case study, human epidermal growth factor (*h*EGF) was selected since it is the most studied growth factor internally at ORF Genetics. Although *h*EGF is not a part of the MESOkine® portfolio, since it is from the human species, the growth factor displays a good model growth factor representative for the purpose of this study. Barley expresses EGF in high yield, and the protein remains stable after processing. The input liquid feed solution containing the extracted *h*EGF, with a native barley matrix that still holds some remaining barley proteins and polysaccharides, was spray dried into a powder form as described below.

### 2.2 Spray drying process

Spray drying trials were executed using next-generation, laboratory-scale Büchi S-300 Advanced Pro Mini Spray Dryer (*Büchi Labortechnik AG, Switzerland).* The spray dryer was coupled with a Büchi S-396 dehumidifier to ensure consistent humidity of the drying air and equipped with a high-performance cyclone to improve collection of the smallest particles. A two-fluid nozzle was used in the trials using air as the drying medium. The spray drying operation was first tested with some feasibility pilot runs.

### *2.2.1 Design of Experiment (DoE)*

DoE is a systematic approach to simultaneously evaluate the effects and interaction of multiple factors that influence the responses of a process. DoE is a component of Quality by Design (QbD), which is a recommended statistical practice in the formulation of drug products by the International Council for Harmonization of Technical Requirements for Pharmaceuticals for Human use (ICH) [17, 36].

Quality characteristics for each product must be pre-determined and must be able to be accurately measured. The measured quality characteristic is called a response. The quality of spray-dried products is influenced by several factors. *Arthur et al.* [37] used DoE to optimize the spray drying process of beer powder and found that moisture content, water activity, solubility, pH, and product yield were the most influential factors. *Ziaee et al.* [38] successfully spray-dried solid dispersions of ibuprofen and used DoE to identify that residual moisture content and particle size were the critical factors for the final yield, but the API/excipient ratio was critical to formulating samples. For our case study, we chose the following factors to investigate and optimize: Feed flow rate (FFR), spray gas, and outlet temperature. Preliminary operation tests assisted in designing the study. The following factors were kept constant between trials: Starting feed solution was identical (constant temperature and product dry weight % w/v), and the drying gas flow rate (aspiration) was kept at maximum capacity.

JMP Pro Version 17.0 (*SAS Institute Inc., Cary, NC*) was used for statistical analysis and model building. We used Response Surface Methodology (RSM) with an *I*-Optimal, custom, DoE design, consisting of 16 runs. RSM is an effective tool to optimize process parameters quickly and efficiently. It can explore the effects of multiple factors simultaneously, investigate factor interactions, and predict the resulting response.

#### *2.2.1.1 Factors and responses*

The FFR, spray gas flow rate, and outlet temperature were defined, in this study, as the potentially critical process parameters, also known as factors in this study. The factors and investigated level ranges are listed in **Table 1.** The levels selected represent the experimental space of the DoE. The selection was based on previous in-house trials and technical restrictions.

The influence of the factors and their interactions resulted in varying powder qualities and characteristics. These effects were measured via the selected output responses: powder output, reconstitution performance, *h*EGF/powder, and total protein/powder to evaluate in this study; see **Table 2**.

| Factor | Level |
|---|---|
| FFR (mL/min) | 4–7 |
| Spray gas (L/h) | 600–1800 |
| Outlet temperature (°C) | 60–100 |

**Table 1.**
*Input factors selected in the DoE custom design used for this case study.*

| Response | Unit of measurement |
|---|---|
| Powder output | % *w/v* |
| Reconstitution score | Scored from 1-4 |
| *h*EGF/powder | µg/mg |
| Total protein/powder | µg/mg |

**Table 2.**
*Output responses selected in the DoE custom design used for this case study.*

### *2.2.1.2 Powder output*

The powder output (% *w/v*) was defined based on the spray-dried powder (g) collected per input liquid solution (mL). The amount of input liquid solution was 50 mL for each run; the exact volume was recorded and considered for the powder output calculations. The vessel collecting the dry powder from each run was weighted before and after the run, to determine the amount of collected powder from the run. For example, if 4.9 g of powder was generated from 50 mL of input liquid, the powder output was determined as 9.7%.

### *2.2.1.3 Reconstitution of powder*

For the reconstitution score assessment, the solid powder sample was dissolved in the solvent (Milli-Q® water) to a dry weight percentage of 8% *w/v*. This value resembled the approximate dry weight (% *w/v*) of the input liquid, when comparing the product output across the samples. This was kept consistent between sample runs by dissolving 200 mg powder in 2.5 mL. Upon reconstitution, the solution was left undisturbed for 15 minutes and then vortexed briefly to obtain a homogenous state. The solution was then visually inspected and ranked based on the solubility performance, the intensity of cloudiness, and color formation. The solubility capability of the dissolved powder was ranked from 1 to 4, from bad (1) to good (4) solubility. Low scores (1) were given to cloudy solutions with visible particles, and high scores (4) were given to solutions that were clear and fully homogenous with no visible particles. Solutions from the reconstitution assessment were used further for the quantifications of the growth factor (see Section 2.2.1.4) and the total protein (see Section 2.2.1.5).

### *2.2.1.4 Growth factor quantification*

For *h*EGF/powder (µg/mg) assessment, the amount of *h*EGF target protein was quantified by capillary-based nano immunoassay, JESS Simple Western™ (*ProteinSimple®, Bio-Techne, Minneapolis, MN, USA*). The *h*EGF amount was quantified within each reconstituted sample based on a generated standard curve of *h*EGF with a known *h*EGF concentration and the concentration of *h*EGF within the powder, which could then be calculated compared to the standard. The *hEGF*/powder (µg/mg) value indicates the product yield and stability of the growth factor after the spray drying step. Receiving a low concentration can, for example, indicate aggregation or irreversible protein denaturation as an indirect estimate, such as the *h*EGF not being able to dissolve after spray drying. Refer to **Appendix A** for a more detailed method description for the growth factor quantification.

#### 2.2.1.5 Total protein quantification

Total protein/powder (μg/mg) was measured with Bradford assay. Since the growth factor is semi-purified, other barley proteins are also present in the final product, and these need to be considered. Therefore, native barley proteins were quantified, along with the recombinant growth factor. Reduction of the total protein within the powder can, for example, indicate that the protein is forming an insoluble material due to the spray drying process. Refer to **Appendix A** for a more detailed method description for the total protein quantification.

### 2.2.2 Optimization: Analysis of DoE data

The data were fit with an RSM model with linear regression (Eq. (1)), and model reduction was performed by enforcing a 95% confidence interval, including only factors and interactions with a p-value <0.05 and factors containing statistically significant effects; refer to **Table 3**. As with the factors, a 95% confidence level was chosen for the responses in the model. The responses in the RSM model that had p-value <0.05 were considered statistically significant; others were disregarded. This RSM model can only predict outcomes for the responses "powder output" and "reconstitution score."

$$Y = a + \sum_{i=1}^{3} b_i X_i + \sum_{i=1}^{3} c_i X_i^2 + \sum_{i=1}^{2} \sum_{j=i+1}^{3} d_{ij} X_i X_j \tag{1}$$

$Y$ is the yield; $a$ is the intercept; $b_i$, $c_i$, and $d_{ij}$ are model coefficients; and $X_i$ and $X_j$ represent the model regressors.

A single sample, *sample #3*, was eliminated from the DoE analysis due to a loss of powder to the cyclone, for technical reasons. However, since this error did not influence the bioactivity of the sample, *sample #3* was included in the following bioactivity measurements and the SEM analysis. All other samples and data were included, and no outliers were detected. Factors and response values are summarized below in **Table 4**.

The variation in the amount of *h*EGF and total protein per powder, respectively, could not be explained by the factors included in the model. This confirms that the spray dryer settings, even at extremities, do not affect either the growth factor quantity or the overall total protein quantity. This data shows the robustness of the spray drying process within the tested ranges.

| Factors and interactions | P value | Responses | P value |
|---|---|---|---|
| Outlet temperature (60, 100) | 0.00057 | Powder output | 0.0010 |
| FFR * Outlet temperature | 0.00169 | Reconstitution score | 0.0029 |
| FFR * Spray gas | 0.00556 | *h*EGF/powder | 0.7066 |
| Spray gas * Spray gas | 0.00654 | Total protein/powder | 0.4137 |
| Spray gas (600, 1800) | 0.00730 ^ | — | — |
| Outlet temperature * Outlet temperature | 0.01846 | — | — |
| FFR (4, 7) | 0.05531 ^ | — | — |

*Outlet temperature (°C), Spray gas (L/h), and FFR (mL/min). ^ denotes factors with containing effects above them.*

**Table 3.**
*Effect summary. Factors and interactions that influence the responses in the RSM model and their corresponding P values.*

| Factors | | | | Responses | | | |
|---|---|---|---|---|---|---|---|
| Sample run | FFR (mL/min) | Spray gas (L/h) | Outlet temperature | EGF/powder (μg/mg) | Total Protein/powder (μg/mg) | Powder output (% *w/v*) | Reconstitution (score) |
| 1 | 4 | 1800 | 60 | 8.31 | 408.1 | 9.80 | 4 |
| 2 | 7 | 1800 | 100 | 7.99 | 411.8 | 9.68 | 2 |
| 3 | 5 | 1200 | 80 | 8.41 | 386.5 | 7.34 | 2 |
| 4 | 7 | 600 | 60 | 7.57 | 377.6 | 8.29 | 2 |
| 5 | 4 | 600 | 100 | 8.15 | 347.6 | 8.41 | 1 |
| 6 | 4 | 600 | 60 | 6.33 | 379.9 | 8.73 | 3 |
| 7 | 5 | 1200 | 60 | 6.46 | 390.6 | 9.88 | 3 |
| 8 | 5 | 1800 | 80 | 6.93 | 370.6 | 9.96 | 3 |
| 9 | 4 | 1800 | 100 | 7.58 | 370.5 | 9.86 | 2 |
| 10 | 7 | 1200 | 80 | 7.59 | 410.2 | 9.77 | 3 |
| 11 | 5 | 600 | 80 | 7.19 | 326.4 | 9.19 | 3 |
| 12 | 4 | 1236 | 80 | 6.66 | 355.4 | 9.92 | 3 |
| 13 | 7 | 600 | 100 | 7.39 | 331.4 | 9.47 | 2 |
| 14 | 7 | 1800 | 60 | 6.70 | 363.6 | 7.9 | 3 |
| 15 | 5 | 1200 | 100 | 7.06 | 361.0 | 9.79 | 2 |
| 16 | 5 | 1200 | 80 | 6.74 | 375.9 | 9.8 | 3 |

**Table 4.**
*Experimental design matrix, containing input factors and output responses from all sample runs.*

Powder output and reconstitution score were found to be statistically significant, and the model can therefore explain the variation in their responses. The outlet temperature affected reconstitution the most, followed by the interactions of *FFR x outlet temperature* and *FFR x spray gas flow rate*. For the powder output, the interactions of FFR x outlet temperature and *FFR x spray gas flow rate* were the most impactful. **Figure 2a** illustrates factor interactions.

A quadratic effect was observed in the model for both the spray gas flow rate and the outlet temperature. A quadratic effect in a statistical model means that an optimum has been observed in the defined experimental space. This can be visualized by a curvature in their responses, as illustrated in **Figure 2b.** Optimum settings, within the tested range, were found for outlet temperature and spray gas flow rate, as curvature was observed in their responses. The quadratic effect for outlet temperature was only significant for powder output, and the quadratic effect for spray gas was only significant for reconstitution.

The optimized settings of the input factors to maximize the responses in this case study were found to be the following:

**Feed Flow Rate** (**FFR**): 4mL/min
**Spray Gas Flow Rate**: 1400L/h
**Outlet Temperature**: 60°C

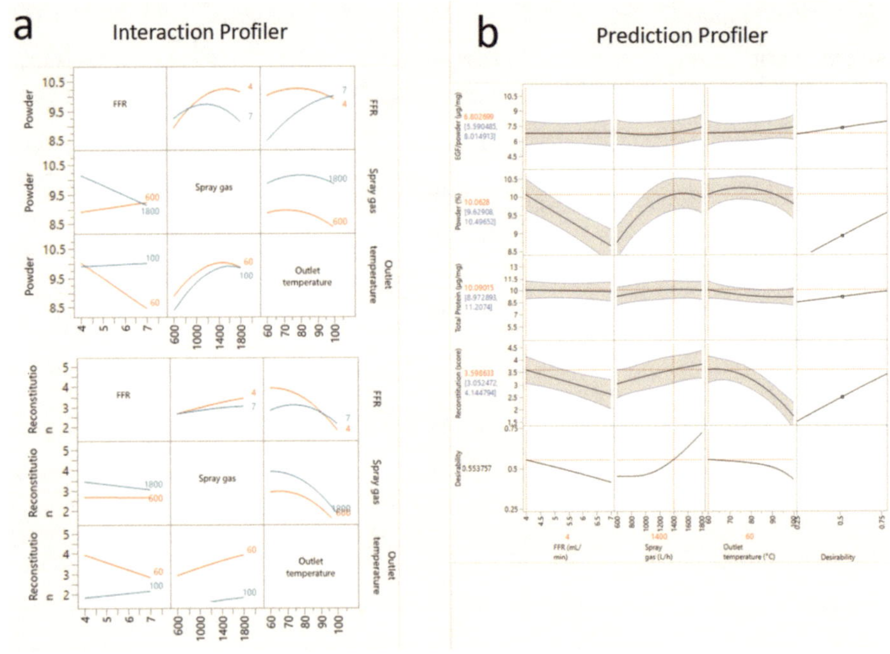


**Figure 2.**
*Effects of the factors on the responses. (a) Interaction profilers for powder and reconstitution show how the different factors interact to affect the responses. (b) Prediction profiler for maximized responses and optimal setting.*

The software calculated the maximized spray gas flow rate settings at 1800 L/h, but we took economic considerations into account when selecting the optimized settings and lowered the selected spray gas flow rate settings to 1400 L/h.

The RSM approach of the DoE in this case study was successful and provided optimum parameter settings for spray drying growth factors in a stabilizing barley matrix with high quality powder and had no impact on the quantity of the target protein or the barley proteins.

### *2.2.3 Particle morphology with scanning electron microscope (SEM)*

Particle morphology analysis was carried out to investigate whether there was a correlation between microscopical particle shapes and the applied spray drying conditions. The surface morphology of the spray-dried particles was examined using a field emission scanning electron microscope (FE-SEM), Supra 25 by Zeiss (*Oberkochen, Germany*). Powder samples were mounted to a sample stub and gold coated. Samples were scanned at a voltage of 3.0 kV, and their images were captured at two magnification levels, 1000× and 5000×.

Selection criteria for the sample runs ultimately taken for SEM characterization were based on analyzing the morphology of the samples expected to have experienced the upper and lower limits for the outlet temperature, reconstitution rating, and particle size.

The analysis of the SEM images served as a qualitative assessment for the given range of parameters in the study. Morphology classification for spray-dried particles as suggested by Prinn et al. [39] can be divided into four categories: (*I*) smooth

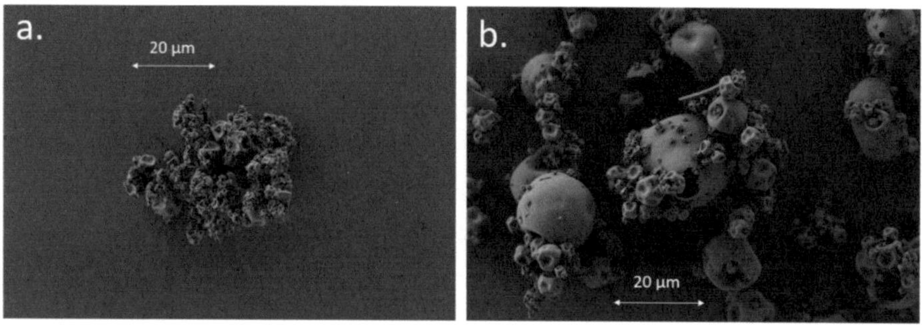


**Figure 3.**
*SEM images of the spray-dried particles at magnification 1000×. Shown are captures from the following samples: (a) Sample #9, expected to have the smallest particles, here shown as several raisin-like particles crumpled together. (b) Sample #4, expected to have the largest particles, here shown as a mixture of larger, smooth spheres; collapsed particles; and wrinkled particles.*

spheres, (*II*) collapsed or dimpled particles, (*III*) wrinkled or raisin-like particles, and (*IV*) highly crumpled or folded structures.

The study design range demonstrated a distribution of most of the above-mentioned different shapes but mostly showed smooth spheres and raisin-like structures. Shown in **Figure 3** are the samples expected to exhibit the smallest particles (sample #9; 4 ml/min, 100°C and 1800 *l/h*) and the largest particles (sample #4; 7 ml/min, 60°C and 600 *l/h*). By comparing these two different particle formations, these reveal only raisin-like particles for the smallest ones (**Figure 3a**), whereas the largest ones show a combination of the two morphologies, smooth spheres and raisin-like particles (**Figure 3b**). The smaller particles seem to tend to clump together. The other tested formulations showed varying amounts of large smooth spheres with smaller particles always forming raisin-like structures with some indications that the higher outlet temperature results in more crumbled structures compared with lower outlet temperatures where smooth surface is more prevalent (data not shown). In other ways, the SEM analysis reveals that particle morphology is not susceptible to changes within the selected range of process parameters.

### *2.2.4 Bioactivity of growth factor*

A cell proliferation assay for *h*EGF using 3 T3 fibroblast cells was performed by SBH sciences (*Natick, MA)* to measure the biological activity of the growth factor for selected samples. The cells were seeded on multi-well plates and incubated with a dilution series of a commercial growth factor standard. After a pre-defined incubation period, cell proliferation or cell death had been measured using a colorimetric assay. The biological activity of *h*EGF was expressed as *ED50* (effective dose), which is the concentration of the growth factor that induces 50% of the maximum assay response. Thus, the lower the *ED50* value, the higher the activity.

The sample selection for the bioactivity analysis is summarized in **Table 5.** This assay was performed to determine whether there was a correlation between biological activity and the other output responses investigated in the case study.

The bioactivity curves of all samples were tested for parallelism to determine whether the samples were statistically different from each other. Parallelism was

| Sample run | Details | FFR (mL/min) | Spray gas (L/h) | Output temperature (°C) | ED50 value (ng/mL) |
|---|---|---|---|---|---|
| 1 | Reconstitution score significantly better than the others | 4 | 1800 | 60 | 0.05–0.08 |
| 2 | Upper extreme conditions for all factors | 7 | 1800 | 100 | 0.08–0.11 |
| 3 | Mid-point of factor levels | 5 | 1200 | 80 | 0.06–0.09 |
| 5 | Reconstitution score significantly worse than the others | 4 | 600 | 100 | 0.05–0.07 |

**Table 5.**
*Sample runs selected for the bioactivity measurement and the resulting ED50 values.*

determined by an F-test using JMP Version 17.0 (*SAS Institute Inc., Cary, NC*) software; see **Appendix B**.

All samples were found to retain their bioactivity after spray drying. Sample #2 was found to have the highest ED50 value, and it was found to be different from the other samples, except for sample #3. Sample #2 had the most extreme settings for all the factors. This indicates that the combination of the extreme process parameters might decrease the bioactivity of spray-dried growth factors. This, however, needs further investigation.

In summary, the biological activity of the growth factor was not disrupted for any of the applied process parameters. Spray drying is a robust process to dry recombinant growth factor solutions while preserving biological activity.

## 3. Continuous, sustainable, and cost-effective process

Spray drying presents clear advantages over freeze drying, especially regarding great scalability capabilities and automation options. It has the potential of making the preservation process more economical.

### 3.1 Continuous system

A continuous process system with a sample bag connection design at a laboratory scale that was set up in-house is shown below in **Figure 4**. The protein liquid in a sterile bag, retrieved straight from the final downstream process filtration, can be fed directly into the spray dryer, surpassing the need for further formulation with the addition of excipients. After the process, the output powder assembles into the collection vessel, which can then be aliquoted into smaller dosages of powder samples for storage.

### 3.2 Economical drying

Reducing the carbon footprint is unquestionably a big industrial focus. Making the shift from freeze drying to spray drying may, in fact, represent a significant step toward reduced energy consumption, thereby lowering the overall climate footprint. *Baeghbali et al.* [40] compared the energy consumption of spray drying and freeze drying, showing that the spray dryer required less than 10% of the energy consumed

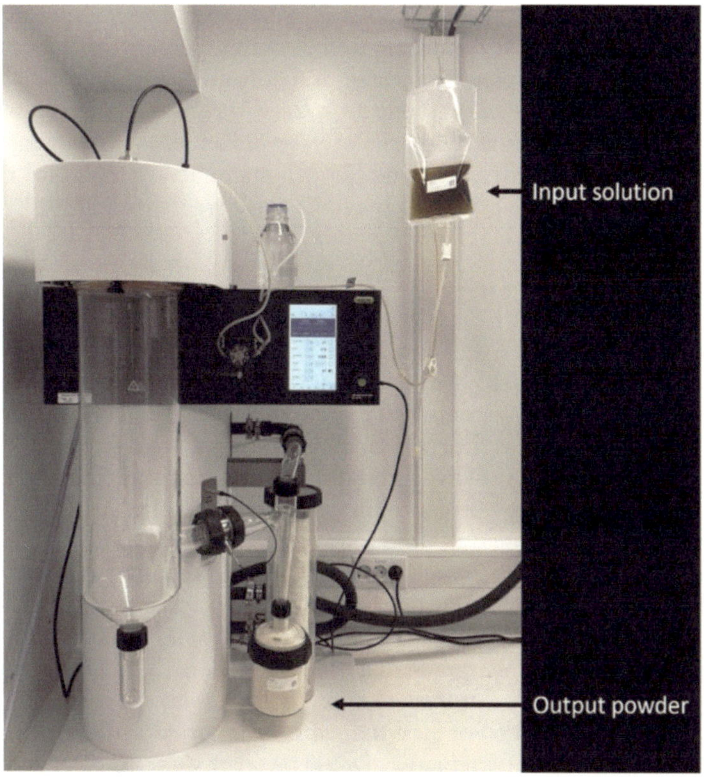


**Figure 4.**
*Continuous spray drying system with sample bag connection design.*

in the freeze drying process, even though noting that the spray drying process was suboptimal and improvements could be made [40].

The DoE approach, in this study, allowed evaluation of the process based on the estimated energy expenditure. Since drying air flow was kept constant throughout the experiments, the aim would be to minimize the inlet air temperature (energy usage per unit time) while maximizing the feed flow rate (reduced processing time) without affecting the quality attributes negatively. Other researchers have shown a strong correlation between the inlet temperature and the outlet temperature [39, 41, 42], whereas they show lesser [39] or even non-significant influence [42] of the feed flow rate on the temperature relationship despite testing much wider feed flow rate ranges than in the current study (3–20 ml/min and 7.3–17.5 ml/min, respectively). This indicates that to reduce the energy consumption in the process, a design space should be created with low outlet temperature and high feed flow rate. In the current study, the best results were achieved at low outlet temperature, which supports reduced energy consumption, whereas the optimized low feed flow rate increases the energy expenditure. The strongest drive in the model to keep feed flow rate at low levels comes, however, from the sharp drop in powder output at higher feed flow rates, meaning a loss in yield at higher feed flow rates. Therefore, considering that the spray drying operation is the last step in the manufacturing process, a drop in yield results in wasted energy and resources in all manufacturing steps upstream from the spray dryer, hence justifying the use of lower feed flow rate for overall reduction in energy consumption.

To evaluate the energy usage of the process, a compact energy meter, Energy-230 Micro LCD (*Vemer, Italy*), was connected to the spray dryer and the dehumidifier. This equipment is designed to display the consumption of active energy in a single-phase system. The electricity usage was roughly 10 kWh when drying one liter of *h*EGF solution. This is consistent with the results of *Baeghbali et al.* [40], showing that a lab-scale spray dryer coupled with a dehumidifier can manufacture high-quality, dry *h*EGF powder with only a fraction of the energy required for freeze drying [40].

### 3.3 Future considerations

Moisture content in spray-dried powder can influence the overall product stability. Higher feed flow rates potentially result in undesirably high water content. Furthermore, the glass transition temperature has been linked with feed flow rate in similar manners as the residual moisture acts as a plasticizer, increasing molecular mobility at lower storage temperatures [43]. Therefore, lower moisture content in dry powders exhibits better long-term stability for protein [44]. Further investigation is needed toward evaluating the powder quality, including comparison of the moisture content, which shall be considered as an added critical output response in future studies. Also, in later strategies and assessments, a long-term stability study of the storage capabilities of the powder should be investigated.

Spray drying is ideal for food-grade material production, like animal-derived recombinant growth factors that are a crucial component of the serum-free media for CCM production. Since cost reduction of growth factors is important for the ultimate success of cell-cultured meat [45], the production should be tailored to this given industry. Food-grade media with lowered associated costs that could still maintain cell proliferation and differentiation at a larger scale would be considered as a success. Later, translation of this application to pilot scale and then eventually to industrial scale will be needed to ensure feasibility to hand the process over to a larger spray drying production.

## 4. Conclusions

- The novelty of this study was to illustrate that the stabilizing effects, generally obtained from excipients, are suggested to be already present from the native barley background matrix, which is a part of the final product.

- The growth factor tolerates high outlet temperatures while retaining its stability in the powder and bioactivity and showing limited effects on the morphology scale.

- The amount of powder collected from the process, along with the ease of powder reconstitution after processing, had the largest effects on the resulting optimization model.

- The spray drying process in this study appears robust enough to surpass the need to add external excipients to the semi-purified growth factor formulation.

- This study demonstrates the feasibility of spray drying bioactive recombinant growth factors embedded in native barley matrix at a laboratory scale.

## Acknowledgements

Thanks go Dr. Raphael Nir (SBH Sciences, Natick, MA) for biological activity measurements in cell proliferation assay.

Thanks go to Tæknisetrið, Reykjavik, Iceland for operational assistance and access to their field emission scanning electron microscope (FE-SEM) for the particle surface morphology characterization.

## Conflict of interest

The authors declare the following interests: affiliation with ORF Genetics, either through employment[1] or paid consultancy services[2].

## Notes

ORF Genetics owns a patent on the barley expression system, ORFEUS™. For this reason, confidential information cannot be disclosed relating to the exact purification procedures and full LC/MS data of the commercial products under the portfolio MESOkine.

## Abbreviations

| | |
|---|---|
| CCM | Cell-cultured meat |
| cGMP | Current good manufacturing practice |
| DoE | Design of Experiments |
| EGF | Epidermal growth factor |
| FFR | Feed flow rate |
| FGF-b | basic Fibroblast growth factor basic |
| GRAS | Generally regarded as safe |
| GFs | Growth factor |
| ICH | International Council for Harmonization of Technical Requirements for Pharmaceuticals for Human use |
| IGF-1 | Insulin-like growth factor 1 |
| QbD | Quality by Design |
| RSM | Response Surface Methodology |
| SD | Spray drying |
| w/v | Weight per volume |

## Appendix A – Materials and Methods in Case Study

## Jess Simple Western

JESS Simple Western™ instrument (ProteinSimple®, Bio-Techne, Minneapolis, MN, USA) is a fully automated and capillary-based system that performs all downstream steps of sample preparation [46]. For preparation of the separation module a 2–40 kDa separation module from ProteinSimple was used which includes a capillary cartridge, pre-filled microplates, wash buffer, 10× sample buffer, lyophilized fluorescent 5x master mix, lyophilized DTT and a lyophilized biotinylated ladder. Standard pack reagents were prepared per manufacturer recommendations, for the 400 mM DTT solution preparation, the fluorescent 5X master mix and the ladder. The samples and *h*EGF standard were prepared. The samples and the standard *h*EGF, with a known concentration, were serially diluted down to 3–104 ng/mL using 0.1× sample buffer. Then, fluorescent 5X master mix was added to all measured samples, so the final mixture is 1 part master mix and 4 parts sample. The samples and standards were heated at 95°C for 5 minutes. The samples were then mixed again by vortex and then spun down and stored on ice. For detection, an anti-rabbit detection module from ProteinSimple was used which includes anti-rabbit secondary antibody, Luminol S, Peroxide, Streptavidin-HRP and antibody diluent. For preparation of the primary antibody, polyclonal rabbit anti-*h*EGF antibody (ab9695, Abcam) was diluted in 1:50 ratio with antibody diluent. For the preparation of the chemiluminescence substrate, 200 µL of luminol-S and 200 µL of peroxide was mixed in a microcentrifuge tube. Then the prepared samples, the biotinylated ladder, antibody diluent for washing and blocking, the diluted primary antibody, the secondary antibody, Streptavidin-HRP for biotinylated ladder capillary only, and luminol-peroxide mix were pipetted into the microplate. The plate was spun down at 1000 ×g for 5 minutes followed by addition of wash buffer and all air bubbles removed with ethanol vapor and the plate placed into Jess and the right program selected in Compass Simple Western ™ software.

### Bradford assay

A Bradford assay was performed to determine total protein concentrations. For each sample, standard or blank (Milli-Q water), 10 µL was pipetted into a microplate well in triplicate. Into each microplate well, 300 µL of Bradford reagent from Thermo Fisher was added and mixed for 30 seconds on a shaker and then incubated for 10 minutes at room temperature. Absorption was measured at 595 nm with a microplate reader (*Thermo Fisher Multi-scan FC*). The average from the triplicate Blank was then subtracted from all other measurements and a 4PL standard curve generated from a pre-diluted Pierce™ Bovine Gamma Globulin (BGG) from Thermo Fisher with a concentration range of 125 µg/mL – 2000 µg/mL. The standard curve is then used to determine the total protein concentration of each unknown sample.

## Appendix B – Design of Experiment (DoE) Results

### Parallelism test in bioassay curves

To test for Parallelism in the Bioassay results, curves were plotted from the Net OD at 490 nm against the log of the concentration (ng/mL). The curves were fitted with a Logistic 4PL fit and a Parallelism test was run between all the possible combinations of curves. A Parallelism F-test where the Prob > F was ≤0.05 means there is no parallelism and therefore the samples are statistically different from each other.

### Bioactivity

See below analysis for bioactivity curves of all samples (**Figure B.1**), raw data of the bioactivity results (**Table B.1**) and results of statistical analysis on bioactivity curves (**Table B.2**).

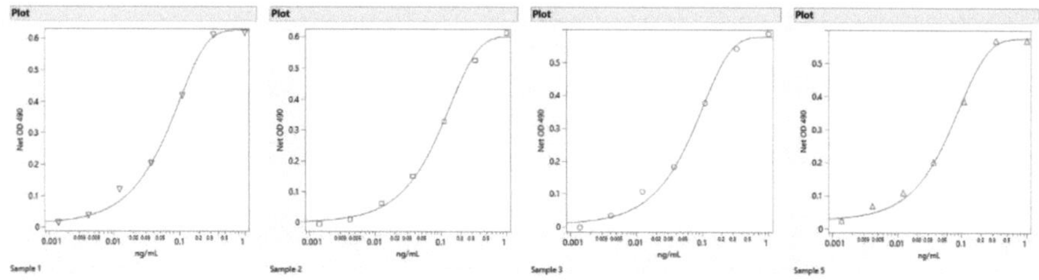


**Figure B.1.**
*Bioactivity curves of all samples.*

| | Net O.D. (ng/ml) | | | | |
|---|---|---|---|---|---|
| **Sample** | **1** | **2** | **3** | **4** | **5** |
| | 1.000 | 0.618 | 0.614 | 0.587 | 0.568 |
| | 0.333 | 0.611 | 0.526 | 0.542 | 0.568 |
| | 0.111 | 0.419 | 0.329 | 0.376 | 0.385 |
| | 0.037 | 0.204 | 0.150 | 0.181 | 0.202 |
| | 0.012 | 0.120 | 0.062 | 0.106 | 0.109 |
| | 0.004 | 0.038 | 0.011 | 0.033 | 0.069 |
| | 0.001 | 0.014 | −0.005 | −0.003 | 0.024 |
| | 0.000 | 0.008 | −0.017 | −0.004 | 0.024 |
| | 0.000 | 0.003 | −0.009 | −0.005 | 0.029 |
| | 0.000 | 0.000 | 0.020 | 0.012 | 0.000 |

**Table B.1.**
*Raw data of the bioactivity results.*

| Sample | Prob > F | Difference |
|---|---|---|
| 3 + 5 | 0.9917 | No |
| 3 + 1 | 0.0571 | No |
| 3 + 2 | 0.1564 | No |
| 5 + 1 | 0.0664 | No |
| 5 + 2 | 0.0205 | Yes |
| 1 + 2 | 0.0017 | Yes |
| 3 + 5 + 1 | 0.1 | No |
| 3 + 5 + 2 | 0.1953 | No |
| 3 + 1 + 2 | 0.0062 | Yes |
| 1 + 5 + 2 | 0.0042 | Yes |
| 3 + 5 + 1 + 2 | 0.0103 | Yes |

**Table B.2.**
*Results of statistical analysis on bioactivity curves.*

**Author details**

Natalía Reynisdóttir[1], Páll Thor Ingvarsson[2], Ásta María Einarsdóttir[1], Arnór Freyr Ingunnarson[1] and Ildikó Nagy[1]*

1 ORF Genetics, Kópavogur, Iceland

2 Faculty of Pharmaceutical Sciences, School of Health Sciences, University of Iceland, Reykjavik, Iceland

*Address all correspondence to: ildiko.nagy.phd@gmail.com

## References

[1] Shanmugaraj B, Bulaon CJ, Phoolcharoen W. Plant molecular farming: A viable platform for recombinant biopharmaceutical production. Plants. 2020;**9**(7):842

[2] Lau OS, Sun SSM. Plant seeds as bioreactors for recombinant protein production. Biotechnology Advances. 2009;**27**(6):1015-1022

[3] Schillberg S, Raven N, Spiegel H, Rasche S, Buntru M. Critical analysis of the commercial potential of plants for the production of recombinant proteins. Front Plant Science. 2019;**2019**:10. Available from: https://www.frontiersin.org/articles/10.3389/fpls.2019.00720

[4] Thomas DR, Penney CA, Majumder A, Walmsley AM. Evolution of plant-made pharmaceuticals. International Journal of Molecular Sciences. 2011;**12**(5):3220-3236

[5] Magnusdottir A, Vidarsson H, Björnsson JM, Örvar BL. Barley grains for the production of endotoxin-free growth factors. Trends in Biotechnology. 2013;**31**(10):572-580

[6] Buyel JF. Plant molecular farming – Integration and exploitation of side streams to achieve sustainable biomanufacturing. Frontiers in Plant Science. 2019;**9**:1893

[7] Emami F, Keihan Shokooh M, Mostafavi Yazdi SJ. Recent progress in drying technologies for improving the stability and delivery efficiency of biopharmaceuticals. Journal of Pharmaceutical Investigation. 2023; **53**(1):35-57

[8] Chen Y, Mutukuri TT, Wilson NE, Zhou Q. Pharmaceutical protein solids: Drying technology, solid-state characterization and stability. Advanced Drug Delivery Reviews. 2021; **172**:211-233

[9] Santos D, Maurício AC, Sencadas V, Santos JD, Fernandes MH, Gomes PS, et al. Spray drying: An overview. In: Biomaterials – Physics and Chemistry – New Edition. London, UK: IntechOpen; 2017. Available from: https://www.intechopen.com/chapters/58222

[10] Sollohub K, Cal K. Spray drying technique: II. Current applications in pharmaceutical technology. Journal of Pharmaceutical Sciences. 2010;**99**(2): 587-597

[11] Spray_Drying_Basic_theory_applications.pdf [Internet]. [cited 2023 Jun 15]. Available from: https://assets.buchi.com/image/upload/v1605790951/pdf/Application-Guides/Spray_Drying_Basic_theory_applications.pdf

[12] Cal K, Sollohub K. Spray drying technique. I: Hardware and process parameters. Journal of Pharmaceutical Sciences. 2010;**99**(2):575-586

[13] Abdul-Fattah AM, Kalonia DS, Pikal MJ. The challenge of drying method selection for protein pharmaceuticals: Product quality implications. Journal of Pharmaceutical Sciences. 2007;**96**(8):1886-1916

[14] Langford A, Bhatnagar B, Walters R, Tchessalov S, Ohtake S. Drying technologies for biopharmaceutical applications: Recent developments and future direction. Drying Technology. 2018;**36**(6):677-684

[15] Pinto JT, Faulhammer E, Dieplinger J, Dekner M, Makert C, Nieder M, et al. Progress in spray-drying

of protein pharmaceuticals: Literature analysis of trends in formulation and process attributes. Drying Technology. 2021;**39**(11):1415-1446

[16] Rezvankhah A, Emam-Djomeh Z, Askari G. Encapsulation and delivery of bioactive compounds using spray and freeze-drying techniques: A review. Drying Technology. 2020;**38**(1–2): 235-258

[17] Ziaee A, Albadarin AB, Padrela L, Femmer T, O'Reilly E, Walker G. Spray drying of pharmaceuticals and biopharmaceuticals: Critical parameters and experimental process optimization approaches. European Journal of Pharmaceutical Sciences. 2019;**127**: 300-318

[18] Ameri M, Maa YF. Spray drying of biopharmaceuticals: Stability and process considerations. Drying Technology. 2006;**24**(6):763-768

[19] Wu J, Wu L, Wan F, Rantanen J, Cun D, Yang M. Effect of thermal and shear stresses in the spray drying process on the stability of siRNA dry powders. International Journal of Pharmaceutics. 2019;**566**:32-39

[20] Wang W. Lyophilization and development of solid protein pharmaceuticals. International Journal of Pharmaceutics. 2000;**203**(1–2):1-60

[21] Crowe JH, Carpenter JF, Crowe LM. The role of vitrification in anhydrobiosis. Annual Review of Physiology. 1998;**60**: 73-103

[22] Costantino HR, Carrasquillo KG, Cordero RA, Mumenthaler M, Hsu CC, Griebenow K. Effect of excipients on the stability and structure of lyophilized recombinant human growth hormone. Journal of Pharmaceutical Sciences. 1998;**87**(11):1412-1420

[23] Allison SD, Chang B, Randolph TW, Carpenter JF. Hydrogen bonding between sugar and protein is responsible for inhibition of dehydration-induced protein unfolding. Archives of Biochemistry and Biophysics. 1999; **365**(2):289-298

[24] Lee G. Spray-drying of proteins. In: Carpenter JF, Manning MC, editors. Rational Design of Stable Protein Formulations: Theory and Practice. Boston, MA: Springer US; 2002. pp. 135-158. DOI: 10.1007/978-1-4615-0557-0_6

[25] Carpenter JF, Pikal MJ, Chang BS, Randolph TW. Rational design of stable lyophilized protein formulations: Some practical advice. Pharmaceutical Research. 1997;**14**(8):969-975

[26] Chen Y, Ling J, Li M, Su Y, Arte KS, Mutukuri TT, et al. Understanding the impact of protein–excipient interactions on physical stability of spray-dried protein solids. Molecular Pharmaceutics. 2021;**18**(7):2657-2668

[27] Landström K, Alsins J, Bergenståhl B. Competitive protein adsorption between bovine serum albumin and β-lactoglobulin during spray-drying. Food Hydrocolloids. 2000;**14**(1):75-82

[28] Wang R, Tian Z, Chen L. A novel process for microencapsulation of fish oil with barley protein. Food Research International. 2011;**44**(9):2735-2741

[29] Meira ACFDO, Morais LCD, Figueiredo JDA, Veríssimo LAA, Botrel DA, Resende JVD. Microencapsulation of β-carotene using barley residue proteins from beer waste as coating material. Journal of Microencapsulation. 2023;**40**(3):171-185

[30] Ibrahim BM, Jun SW, Lee MY, Kang SH, Yeo Y. Development of

inhalable dry powder formulation of basic fibroblast growth factor. International Journal of Pharmaceutics. 2010;**385**(1):66-72

[31] Schultz I, Vollmers F, Lühmann T, Rybak JC, Wittmann R, Stank K, et al. Pulmonary insulin-like growth factor I delivery from trehalose and silk-fibroin microparticles. ACS Biomaterials Science & Engineering. 2015;**1**(1):119-129

[32] Andreopoulos FM, Persaud I. Delivery of basic fibroblast growth factor (bFGF) from photoresponsive hydrogel scaffolds. Biomaterials. 2006; **27**(11):2468-2476

[33] Zhang M, Liu J, Li M, Zhang S, Lu Y, Liang Y, et al. Insulin-like growth factor 1/insulin-like growth factor 1 receptor signaling protects against cell apoptosis through the PI3K/AKT pathway in glioblastoma cells. Experimental and Therapeutic Medicine. 2018;**16**(2): 1477-1482

[34] Vilatte A, Spencer-Milnes X, Jackson HO, Purton S, Parker B. Spray drying is a viable technology for the preservation of recombinant proteins in microalgae. Microorganisms. 2023;**11**(2): 512

[35] Hussain A, Ali S, Hussain A, Hussain Z, Manzoor MF, Hussain A, et al. Compositional profile of barley landlines grown in different regions of Gilgit-Baltistan. Food Science & Nutrition. 2021;**9**(5):2605-2611

[36] Tietje C, Brouder A, editors. International Conference on Harmonisation of Technical Requirements for Registration of Pharmaceuticals for Human Use. In: Handbook of Transnational Economic Governance Regimes. Nijhoff: Brill; 2010. pp. 1041-1053. Available from: https://brill.com/view/book/edcoll/9789004181564/Bej.9789004163300.i-1081_085.xml

[37] Arthur DA, Xiaofeng R. Optimization of spray drying conditions for production of barley beer powder. International Journal of Science and Research (IJSR). 2022;**11**(4):1132-1139

[38] Ziaee A, Albadarin AB, Padrela L, Faucher A, O'Reilly E, Walker G. Spray drying ternary amorphous solid dispersions of ibuprofen – An investigation into critical formulation and processing parameters. European Journal of Pharmaceutics and Biopharmaceutics. 2017;**120**:43-51

[39] Prinn KB, Costantino HR, Tracy M. Statistical Modeling of protein spray drying at the lab scale. AAPS PharmSciTech. 2002;**3**(1):4

[40] Baeghbali V, Niakousari M, Farahnaky A. Refractance window drying of pomegranate juice: Quality retention and energy efficiency. LWT – Food Science and Technology. 2016;**66**: 34-40

[41] Maa YF, Costantino HR, Nguyen PA, Hsu CC. The effect of operating and formulation variables on the morphology of spray-dried protein particles. Pharmaceutical Development and Technology. 1997;**2**(3):213-223

[42] Maltesen MJ, Bjerregaard S, Hovgaard L, Havelund S, van de Weert M. Quality by design – Spray drying of insulin intended for inhalation. European Journal of Pharmaceutics and Biopharmaceutics. 2008;**70**(3):828-838

[43] Alves NN, Messaoud GB, Desobry S, Costa JMC, Rodrigues S. Effect of drying technique and feed flow rate on bacterial survival and physicochemical properties of a non-dairy fermented probiotic juice

powder. Journal of Food Engineering. 2016;**189**:45-54

[44] Maa YF, Nguyen P, Andya J, Dasovich N, Sweeney T, Shire S, et al. Effect of spray drying and subsequent processing on residual moisture content and physical/biochemical stability of protein inhalation powders. Pharmaceutical Research. 1998;**15**: 768-775

[45] O'Neill EN, Cosenza ZA, Baar K, Block DE. Considerations for the development of cost-effective cell culture media for cultivated meat production. Comprehensive Reviews in Food Science and Food Safety. 2021; **20**(1):686-709

[46] Sormunen A, Koivulehto E, Alitalo K, Saksela K, Laham-Karam N, Ylä-Herttuala S. Comparison of automated and traditional Western blotting methods. Methods in Protocol. 2023;**6**(2):43